**Advanced Courses in Mathematics
CRM Barcelona**

Centre de Recerca Matemàtica

Managing Editor:
Enric Ventura

More information about this series at http://www.springer.com/series/5038

Valery Alexeev

Moduli of Weighted Hyperplane Arrangements

Editors for this volume:
Gilberto Bini, Università degli Studi di Milano
Martí Lahoz, Université Paris Diderot - Paris 7
Emanuele Macrì, Northeastern University
Paolo Stellari, Università degli Studi di Milano

 Birkhäuser

Valery Alexeev
Department of Mathematics
University of Georgia
Athens, GA, USA

ISSN 2297-0304   ISSN 2297-0312   (electronic)
Advanced Courses in Mathematics - CRM Barcelona
ISBN 978-3-0348-0914-6   ISBN 978-3-0348-0915-3   (eBook)
DOI 10.1007/978-3-0348-0915-3

Library of Congress Control Number: 2015940007

Mathematics Subject Classification (2010): Primary: 14D06; Secondary: 52B40

Springer Basel Heidelberg New York Dordrecht London
© Springer Basel 2015

Printed on acid-free paper

Springer Basel AG is part of Springer Science+Business Media (www.birkhauser-science.com)

Contents

Preface

These notes come from the lectures delivered during the conference "Compactifying Moduli Spaces", held in May 2013 at the Centre de Recerca Matemàtica (CRM) in Barcelona. In them, the speakers report on recent research on moduli theory, from different points of view.

In recent years, moduli spaces have been investigated for their diverse applications to Algebraic Geometry, Number Theory, String Theory, and Quantum Field Theory, just to mention a few. In particular, the notion of compactification of moduli spaces has been crucial for solving various open problems and long-standing conjectures.

The compactification problem can be approached by various techniques. Geometric Invariant Theory, Hodge Theory and the Minimal Model Program (MMP) come into play as different approaches to construct and compactify moduli spaces. All these approaches shed light on particular aspects of moduli problems.

In this volume, Valery Alexeev reports on compactification of moduli spaces in a large class where computations are possible, namely that of weighted stable hyperplane arrangements (shas).

It is our hope that these lectures notes will illustrate the wide and rich variety of ideas and theories that were developed with the general aim of understanding moduli spaces and their geometry.

Gilberto Bini
Martí Lahoz
Emanuele Macrì
Paolo Stellari

Introduction

These notes were written to complement my talks in the workshop "Compactifying Moduli Spaces" at Centre de Recerca Matemàtica, Barcelona, May 27 to 31, 2013. They concern moduli spaces of higher-dimensional stable pairs.

Stable surfaces and their compact moduli were introduced by Kollár and Shepherd-Barron in [42]; an extension to pairs, and also to stable maps was given in [2, 3]. Many crucial pieces, from the correct way of posing the moduli question to the numerous technical questions, were filled in later by many people.

These days, a number of introductions into this theory are available, e.g., [39, 41]. The aim of these notes is *not* to provide yet another general introduction. The focus is narrower. The ratio of the number of papers on the one-dimensional stable curves versus the higher-dimensional case is at least 100 to 1. Some of the reasons are obvious; for instance, the one-dimensional case obviously presents for less technical difficulties. But the main reason is that the one-dimensional case is so much more amenable to combinatorial methods. In contrast, a large part of the higher-dimensional theory consists of pure existence theorems. Concrete computations are so much harder to perform, and there are few completely computed cases.

One large class where explicit computations *are* possible is the class of weighted stable hyperplane arrangements (*shas*). They provide compactifications for the moduli spaces of log canonical hyperplane arrangements $(\mathbb{P}^{r-1}, \sum_{i=1}^{n} b_i B_i)$. As applications, one also obtains various results about moduli spaces of surfaces of general type and K3 surfaces, typically by considering Galois covers of $\mathbb{P}^2$ ramified in special configurations of lines.

I will try to explain, as concretely as possible, how to work with such weighted stable hyperplane arrangements, and how to make computations about them and their moduli spaces. The whole story is an intricate interplay of Minimal Model Program, Geometric Invariant Theory, Matroid Theory, and polytopal tilings. It is my hope that this concrete introduction will allow more people to enter this research field.

Another reason to concentrate on weighted shas is their applicability to other cases. These include the idea of complementary degenerations, and the idea that a stable pair should correspond to some kind of polytopal or almost polytopal tiling with "integral" vertices. The geography of the moduli spaces (as the coefficients b_i change) is also expected to be similar in the general case.

The present notes should serve as a supplement to the research papers [7, 13] and to the earlier papers [4, 9, 27, 33]. We do not repeat any proofs. Instead, we introduce the necessary combinatorial tools, state the theorems, and go in detail through some illuminating examples. One can think of these notes as a pictorial introduction to the above papers. We also provide results of some computer-aided computations.

The content is as follows. In Chapter 1 we give a quick and rather superficial introduction to the general theory of stable pairs and their moduli spaces, and state what is known. In Chapter 2 we explain the theory of stable toric varieties and pairs. We finish the chapter by explaining how to reduce the moduli of hyperplane arrangements to the moduli of stable toric varieties. The matroid polytopes make an appearance, motivating the next two chapters. In Chapter 3 we introduce as much matroid theory as necessary for our purposes, with a brief detour into regular matroids (important for degenerations of abelian varieties). Chapter 4 is devoted to matroid polytopes and tilings. This includes partial tilings and "cuts". In Chapter 5 we get to the heart of the theory and state the main results. We also illustrate it in dimensions 1 and 2, giving complete classification for $n \leq 6$ lines. Finally, in Chapter 6 we go through some applications: computations for several classes of surfaces of general type and a special low-dimensional case of K3 surface pairs.

I would like to thank János Kollár for many useful comments on Chapter 1 of these notes. The author would like to acknowledge partial support by the NSF under the grant DMS-1200726.

Chapter 1

Stable Pairs and Their Moduli

1.1 The curve case

The focus of these lectures is the higher-dimensional case, and it is hoped that the reader already has some familiarity with the one-dimensional case. So we will be rather brief.

Definition 1.1.1. Fix n real numbers $0 < b_i \leq 1$. A *weighted stable curve* for the weight $\boldsymbol{b} = (b_1, \ldots, b_n)$ is a pair $(X, B = \sum b_i B_i)$ of a reduced connected projective curve X together with n points $B_i \in X$ such that:

1. (Singularities) X has at worst double normal crossings as singularities (locally analytically isomorphic to $xy = 0$). The points B_i may coincide, but they should be different from the nodes, and the sum of the weights should satisfy $\mathrm{mult}_x(B) = \sum_{B_i = x} b_i \leq 1$ for any point $x \in X$.
2. (Numerical) The $\mathbb{R}$-divisor $K_X + B$ is ample.

Here K_X denotes the dualizing sheaf ω_X, which is an invertible sheaf on a nodal curve. The numerical condition is equivalent to saying that for any irreducible component $E \subset X$ the degree of the restriction $(K_X + B)|_E$ is positive:

$$\deg(K_X + B)|_E = 2p_a(E) - 2 + E.(X - E) + \sum_{B_i \in E} b_i.$$

Here, we used the adjunction formula $K_X|_E = K_E(X - E)$, and the formula $\deg K_E = 2p_a(E) - 2$.

This degree is automatically positive if either $p_a(E) \geq 2$ or $p_a(E) = 1$ and $E.(X - E) > 0$. Thus, for a curve X of arithmetic genus $g := p_a(X) \geq 2$, the only condition is for the irreducible curves $E \simeq \mathbb{P}^1$, and it is

$$E.(X - E) + \sum_{B_i \in X} b_i > 2.$$

Thus, we are adding the weights of "special" points on E, and the points of intersection of E with the rest of X count with weight 1.

Example 1.1.2. If all the weights are $b_i = 1$, then the points B_i must be distinct, and on every component $\simeq \mathbb{P}^1$ there should be at least three special points. Thus, $(X, B_1, \ldots, B_n)$ is an ordinary Deligne–Mumford–Knudsen n-pointed stable curve.

The following theorem of Hassett [30] generalizes Deligne–Mumford [17] and Mumford–Knudsen to the case of arbitrary weights:

Theorem 1.1.3. *For any n, $\boldsymbol{b}$ and $g \geq 0$ the moduli stack $\overline{\mathcal{M}}_{g,b}$ of weighted stable curves of arithmetic genus g is a smooth Deligne–Mumford stack with a projective moduli space $\overline{\mathrm{M}}_{g,b}$.*

In the case $g = 0$, the moduli space is fine, and we can identify $\overline{\mathcal{M}}_{0,b}$ with the projective scheme $\overline{\mathrm{M}}_{0,b}$. For $g \geq 1$, it is a coarse moduli space.

We denote by $(\mathcal{X}, \mathcal{B}_1, \ldots, \mathcal{B}_n) \to \overline{\mathcal{M}}_{g,b}$ the universal family over the moduli stack.

The next question is that of "geography": how do the spaces $M_{g,n}$ change when the weight $\boldsymbol{b}$ changes?

Definition 1.1.4. The weight domain $\mathcal{D}_g(n)$ is defined to be the set $\{\boldsymbol{b} \in (0,1]^n\}$. For genus $g = 0$ we additionally require that $\sum b_i > 2$, in order to have $\deg(K_{\mathbb{P}^1} + \sum b_i B_i) > 0$.

Definition 1.1.5. A chamber decomposition of $\mathcal{D}_g(n)$ into locally closed strata is obtained by cutting it by the hyperplanes $\boldsymbol{b}(I) = 1$ for all subsets $I \subset \overline{n}$.

Here, we adopt the notation $\boldsymbol{b}(I) := \sum_{i \in I} b_i$ and $\overline{n} = \{1, \ldots, n\}$.

We also say that $(b'_1, \ldots, b'_n) \geq (b_1, \ldots, b_n)$ if $b'_i \geq b_i$ for all $i = 1, \ldots, n$.

Theorem 1.1.6 ([30]). *The following holds:*

1. *(Same chamber) For $\boldsymbol{b}, \boldsymbol{b}'$ in the same locally closed chamber, the moduli stacks are the same and the universal families $\mathcal{X} \to \overline{\mathcal{M}}$ are the same.*

2. *(Specialization from above) For $\boldsymbol{b}' \in \overline{\mathrm{Ch}(\boldsymbol{b})}$ (denoted $\boldsymbol{b}' \in \overline{\boldsymbol{b}}$) and $\boldsymbol{b}' \geq \boldsymbol{b}$, there exist contraction morphisms*

$$
\begin{array}{ccc}
\mathcal{X}' & \longrightarrow & \mathcal{X} \\
\downarrow & & \downarrow \\
\overline{\mathcal{M}}_{g,b'} & \longrightarrow & \overline{\mathcal{M}}_{g,b}.
\end{array}
$$

Further, the map on the moduli space is an isomorphism if $|I| = 2$.

3. *(Specialization from below) For $\boldsymbol{b}' \in \overline{\mathrm{Ch}(\boldsymbol{b})}$ and $\boldsymbol{b}' \leq \boldsymbol{b}$, both the moduli spaces and the universal families are the same.*

One can prove that, when crossing one of the walls $\boldsymbol{b}(I) = 1$ *generically* (i.e., no other inequalities $\boldsymbol{b}(J) \leq 1$ change), the bigger moduli stack is the blowup $\overline{\mathcal{M}}' = \mathrm{Bl}_Z \overline{\mathcal{M}}$ of the smaller moduli stack along the smooth substack Z parameterizing the curves where the points B_i with $i \in I$ coincide; see, e.g., [11]. This wall crossing is illustrated in Figure 1.1.

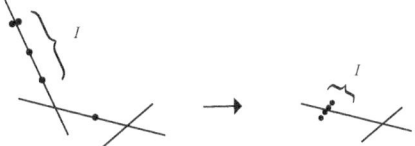

Figure 1.1: Crossing a wall $b(I) = 1$.

1.2 Minimal model program: main definitions and results

We will accept MMP as a black box machine. You feed it a variety or a pair and it spits out a better one. The necessary definitions for singularities (terminal, canonical, log terminal, log canonical, semi-log canonical) will be given later.

1.2.1 MMP machine for varieties

Let us consider the following minimal model program (MMP for short):

Input:
1. A smooth projective variety X.
2. Or, more generally, a normal projective variety X with terminal singularities.

Outputs:
1. Either a minimal model $X_{\min}$ with nef canonical divisor $K_{X_{\min}}$ and terminal singularities (a divisor D is *nef* if $D.C \geq 0$ for any effective curve C), or a Mori–Fano fibration $X' \to Y$ with relatively ample $-K_{X'}$ and $\dim Y < \dim X' = \dim X$.
2. If K_X is big, then also the canonical model X_{can} with ample $K_{X_{\text{can}}}$ and canonical singularities.

So, MMP is a machine for improving the properties of the canonical divisor. What happens in between is really not that important. But here are the important parts:

1. The rational map $X \dashrightarrow X_{\min}$ is birational and it does not create divisors: some divisors may be contracted, but no new divisors are created. For the ranks of Picard groups one has $\rho(X_{\min}) \leq \rho(X)$.

2. The minimal model is usually not unique, but the canonical model is (provided K_X is big). It is obtained from a minimal model by a linear system $|dK_{X_{\min}}|$ for $d \gg 0$. There is also a way to obtain the canonical model directly from X, by the formula $X_{\text{can}} = \operatorname{Proj} R(X, K_X)$, where for any divisor D we set

$$R(X, D) := \bigoplus_{d \geq 0} H^0\big(X, \mathcal{O}(dD)\big).$$

Here, we use the following standard notation. For an integral divisor $D = \sum d_i D_i$ with irreducible D_i on a normal variety X, the divisorial sheaf $\mathcal{O}(D)$ is the $\mathcal{O}_X$-subsheaf of the constant sheaf $\mathcal{K}_X$ of rational functions whose local sections are rational sections with effective $(f) + D$, i.e., $\text{mult}_{D_i}(f) \geq -d_i$. This definition makes perfect sense if $d_i \in \mathbb{R}$ and one has $\mathcal{O}_X(\sum d_i D_i) = \mathcal{O}_X(\lfloor d_i \rfloor D_i)$.

The singular locus of a normal variety has codimension bigger than or equal to 2 and, if $j: U \to X$ is the inclusion of the nonsingular locus, then $\mathcal{O}_X(D) = j_* \mathcal{O}_U(D|_U)$ is the push-forward of an invertible sheaf. The sheaves of this form are called divisorial. In particular, $\mathcal{O}_X(dK_X)$ is a divisorial sheaf on a normal variety for any $d \in \mathbb{Z}$.

The ring $R(X, K_X)$ is called the canonical ring. In the cases where MMP has been proved (listed below) it is a finitely generated ring over the base field k. To have $\dim X_{\text{can}} = \dim X$, the plurigenus $h^0(X, \mathcal{O}(dK_X))$ has to grow as $c \cdot d^{\dim X}$. This is the definition of a *big* divisor.

3. Let us emphasize the following very important point: *birationally isomorphic smooth varieties (or varieties with canonical singularities) have the same canonical model*. Indeed, for smooth varieties and $d \geq 0$, the space $H^0(X, \mathcal{O}(dK_X))$ is a birational invariant. If X is a variety with canonical singularities (see definition below), then $H^0(X, \mathcal{O}(dK_X)) = H^0(Y, \mathcal{O}(dK_Y))$ for any resolution of singularities $Y \to X$.

4. If the starting variety X is $\mathbb{Q}$-factorial, then all intermediate steps and $X_{\min}$ are $\mathbb{Q}$-factorial as well, but X_{can} may not be. (A variety is called $\mathbb{Q}$-factorial if for any Weil divisor some positive multiple of it is Cartier.)

Here is some information about the internals of the machine: in between there is a sequence of birational transformations, $X = X_0 \dashrightarrow X_1 \dashrightarrow \cdots \dashrightarrow X_n$, which are either *divisorial contractions* $X_i \to X_{i+1}$ (contracting a divisor on X_i to a smaller-dimensional subvariety of X_{i+1}), or a *flip*, i.e., a diagram of the form

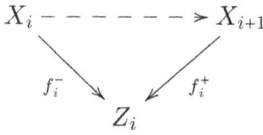

in which f_i^- and f_i^+ are *small contractions*, isomorphisms in codimension 1. (For example, if $\dim X_i = 3$, then both f_i^- and f_i^+ contract some curves.)

1.2.2 MMP machine for pairs

A $\mathbb{Q}$-*divisor* (resp. $\mathbb{R}$-*divisor*) is a formal linear combination $B = \sum b_i B_i$ with $b_i \in \mathbb{Q}$ (resp. $b_i \in \mathbb{R}$) and B_i being effective divisors. Usually, they are assumed to be irreducible and distinct, but we have to omit both of these conditions in order to work with moduli of pairs. Thus, B_i are simply effective $\mathbb{Z}$-divisors, not necessarily irreducible, and they may have components in common.

We require $b_i \geq 0$. For log canonical singularities one will automatically have $b_i \leq 1$. This does not have to be required in advance.

Input:
1. A pair $(X, B = \sum B_i)$ of a smooth projective variety X and a $\mathbb{Q}$-divisor or $\mathbb{R}$-divisor $B = \sum b_i B_i$ such that $\bigcup B_i$ is a normal crossing divisor.
2. Or, more generally, a log canonical pair $(X, B = \sum b_i B_i)$.

Outputs:
1. Either a minimal model $(X_{\min}, B_{\min})$ with nef divisor $K_{X_{\min}} + B_{\min}$ and dlt or log canonical singularities, or a Mori–Fano fibration $X' \to Y$ with relatively ample $-(K_{X'} + B')$ and $\dim Y < \dim X' = \dim X$.
2. If $K_X + B$ is big, then also the log canonical model $(X_{\text{can}}, B_{\text{can}})$ with ample $K_{X_{\text{can}}} + B_{\text{can}}$ and log canonical singularities.

Again, a minimal model is usually not unique, but the log canonical model is, and $X_{\text{can}} = \operatorname{Proj} R(X, K_X + B)$. The question of independence of the log canonical model of X is a little bit more delicate, see Lemma 1.2.4.

1.2.3   Standard singularities

One of the main revelations of the MMP since the earliest days was that, in order to achieve good properties of the canonical class in dimension ≥ 3, one must work with singular varieties. Here are the standard definitions.

Let X be a normal variety and $f: Y \to X$ a resolution of singularities such that the exceptional set $\bigcup E_j$ is a normal crossing divisor. When working with a pair, we additionally assume that $\bigcup f_*^{-1} B_i \cup E_j$ is a normal crossing divisor, where $f_*^{-1} B_i$ is our notation for the strict preimages of B_i. Such a resolution is called a *log resolution*. It exists for characteristic 0 by Hironaka, and in dimension 2 for arbitrary characteristic.

We start with canonical and terminal singularities. Assume that K_X is a $\mathbb{Q}$-Cartier divisor, i.e., that some positive multiple $N K_X$ is a Cartier $\mathbb{Z}$-divisor. Cartier divisors can be pulled back (just pull back the local equation). This gives us the magical formula

$$K_Y \sim_{\mathbb{Q}} f^*(K_X) + \sum a_j E_j.$$

The difference between K_Y and $f^*(K_X)$ consists entirely of the exceptional divisors since they coincide outside of the exceptional locus. The coefficients a_j are called the *discrepancies*. If $N K_X$ is Cartier, then $a_j \in \frac{1}{N}\mathbb{Z}$, so $a_j \in \mathbb{Q}$.

Definition 1.2.1. The singularities of X are called *terminal* if $a_j > 0$ for all j and *canonical* if $a_j \geq 0$ for all j.

It is easy to see that this definition does not depend on the log resolution (see the argument for *klt* and *lc* below). A major consequence of "canonical" is that, for any d divisible by N, one has $H^0(dK_Y) = H^0(dK_X)$. Indeed, the pullbacks of sections of $\mathcal{O}_X(dK_X)$ remain regular. And, since $NK_Y = NK_X+$ (effective exceptional Cartier divisor), there are no new sections.

Now let $(X, B = \sum b_i B_i)$ be a pair with $b_i \geq 0$. The divisor $K_X + B$ is $\mathbb{Q}$-*Cartier* if, for some positive integer N, the coefficients $Nb_i \in \mathbb{Z}$ and $N(K_X + B)$ is a Cartier $\mathbb{Z}$-divisor. One has to be careful about $\mathbb{R}$-divisors:

Definition 1.2.2. An $\mathbb{R}$-*Cartier* divisor is an $\mathbb{R}$-linear combination of Cartier $\mathbb{Z}$-divisors. (Divisors here are sums of irreducible subvarieties, there is nothing in this definition about linear equivalence.) When we say that $K_X + B$ is $\mathbb{R}$-Cartier, this means that for some concrete representative $D \sim K_X$, the divisor $D + B$ is $\mathbb{R}$-Cartier. But then, of course, the same is true for any other representative.

Two $\mathbb{R}$-divisors are $\mathbb{R}$-*linearly equivalent*, written $D_1 \sim_{\mathbb{R}} D_2$, if $D_1 - D_2$ is an $\mathbb{R}$-linear combination $\sum c_i(f_i)$ of principal Cartier divisors.

Since Cartier divisors can be pulled back, we again have the magical formula

$$K_Y \sim_{\mathbb{Q}} f^*\left(K_X + \sum b_i B_i\right) + \sum a_D D,$$

where the sum goes over all irreducible divisors on Y.

Definition 1.2.3. A pair $(X, B = \sum b_i B_i)$ is called *Kawamata log terminal* (klt) if all $a_D > -1$, and *log canonical* (lc) if all $a_D \geq -1$.

If D is not f-exceptional, then the coefficient a_D is $-\sum b_i \operatorname{mult}_{f_*D}(B_i)$. Thus, $a_D > -1$ means $\sum b_i \operatorname{mult}_{f_*D}(B_i) < 1$ (resp. ≤ 1 for lc). If the divisors B_i are irreducible and distinct, then this means simply $b_i < 1$ (resp. $b_i \leq 1$ for lc). But it is enough with B_i's being non-irreducible and having components in common. Then we are saying that, after rewriting $\sum b_i B_i = \sum d_k D_k$ with irreducible distinct D_k's, one must have $d_k < 1$ (resp. $d_k \leq 1$ for lc). For the f-exceptional divisors E_j, the condition says that the discrepancies a_j must satisfy $a_j > -1$ (resp. $a_j \geq -1$ for lc).

The following fact is a basic computation underlying these definitions: if one starts with a reduced normal crossing divisor $\bigcup B_i$ and blows up the intersection of several B_i's then $f^*(K_X + \sum B_i) \sim_{\mathbb{Q}} K_Y + \sum f_*^{-1} B_i + E$, where E is the exceptional divisor of the blowup. So the coefficient 1 is very natural, and the inequalities do not change if one replaces Y by a variety Y' obtained from it by a sequence of blowups (and any other resolution is dominated by such). This shows independence of Y.

There are several flavors of log terminal singularities for pairs: log terminal (lt), divisorially log terminal (dlt), and pure log terminal (plt). They will not be important for us. There is also an important extension of lc to non-normal varieties which we will introduce below: semi-log canonical (slc). Similarly, one can also define slt, sdlt, etc.

1.2.4 Uniqueness of the log canonical model for pairs

Lemma 1.2.4. *Let $(X, B = \sum b_i B_i)$ be an lc pair with a $\mathbb{Q}$-Cartier divisor $K_X + B$ and $f: Y \to X$ be a morphism with exceptional divisors E_j. Then*

$$\operatorname{Proj} R(X, K_X + B) = \operatorname{Proj} R\left(Y, K_Y + f_*^{-1} B + \sum E_j\right).$$

Proof. Indeed, $K_Y + f_*^{-1} B + \sum E_j \sim_{\mathbb{Q}} f^*(K_X + B) + \sum(1 + a_j) E_j$, and the difference $\sum(1 + a_j) E_j$ is effective and f-exceptional, so for all d with $db_i \in \mathbb{Z}$ and $d(K_X + B)$ Cartier, one has $H^0(\mathcal{O}_X(d(K_X + B))) = H^0(\mathcal{O}_Y(d(K_Y + f_*^{-1} B + \sum E_j)))$. $\square$

Next, we are going to give a non-standard definition, not used much (or at all) in the literature.

Definition 1.2.5. Let $(X, B = \sum b_i B_i)$ be a pair with $b_i \geq 0$ (which may or may not be log canonical). Assume that $K_X + B$ is $\mathbb{Q}$-Cartier or $\mathbb{R}$-Cartier. Rewrite $K_X + \sum b_i B_i = \sum d_k D_k$ with distinct irreducible D_k, and let $d_k' = \min(d_k, 1)$. Thus,

$$0 < d_k' = \min\left(\sum b_i \operatorname{mult}_{D_k}(B_i), 1\right) \leq 1.$$

Let $f: Y \to X$ be a log resolution of (X, B) with exceptional divisors E_j. The *log canonical model of (X, B)* is defined to be

$$\operatorname{Proj} R\left(Y, K_Y + \sum d_k' f_*^{-1} D_k + \sum E_j\right),$$

provided that this ring is finitely generated.

Note that the pair $(Y, \sum d_k' f_*^{-1} D_k + \sum E_j)$ is lc. By the above Lemma 1.2.4, this definition does not depend on the choice of a log resolution, since any two such resolutions can always be dominated by a third one.

1.2.5 Relative case

The MMP machine also works in a relative situation, when the input is a projective morphism $\pi: X \to S$ over an arbitrary variety S. The output is a *relative minimal model* with π-nef $K_{X_{\min}}$ (resp. $K_{X_{\min}} + B_{\min}$) or a *relative canonical model* with π-ample $K_{X_{\operatorname{can}}}$ (resp. $K_{X_{\operatorname{can}}} + B_{\operatorname{can}}$) if one starts with a π-big canonical divisor K_X. Further, one has $X_{\operatorname{can}} = \operatorname{Proj}_S R_S(X, K_X + B)$, where

$$R_S(X, K_X + B) = \bigoplus_{d \geq 0} \pi_* \mathcal{O}(d(K_X + B))$$

is a relative canonical ring, an $\mathcal{O}_S$-algebra.

A divisor is π-nef if $D.C \geq 0$ for any curve C collapsed by π, i.e., $\pi(C)$ is a point.

A divisor is π-ample if a positive multiple of it is a pullback of $\mathcal{O}(1)$ from some $\mathbb{P}^n \times S$; if S is projective, this simply means that $D = \operatorname{ample} + \pi^* D'$.

A divisor is π-big if its restriction to a generic fiber is big. For a birational morphism, this is an empty condition.

An alternative name for π-nef is nef *over* S (and, similarly for the other notions).

1.2.6   When is MMP known to be true?

The MMP is still a conjecture in the general case. It is currently known only in dimension 3 for characteristic 0, and in dimension 2 for any characteristic (where it is fairly easy).

A huge step in MMP for arbitrary dimension was made in [15], but this is only for klt pairs. We do need coefficients $b_i = 1$, however, for example to handle varieties with slc singularities.

1.3   Minimal model program and one-parameter degenerations

Here is the essential and most basic application of MMP to the complete moduli of algebraic varieties. For degenerations of curves this was used by Shafarevich [55] and Deligne–Mumford [17]. Kollár and Shepherd-Barron realized in [42] that the same construction can be applied to degenerations of surfaces.

Let $\pi^0\colon X^0 \to S^0$ be a family over a punctured curve $S^0 = S \smallsetminus \{0\}$. We want to extend it to a nice complete family over S, perhaps after a finite base change $S' \to S$, $X^0 \times_S S' \to (S')^0$, which is a standard thing to do when working with moduli of varieties with finite automorphism groups.

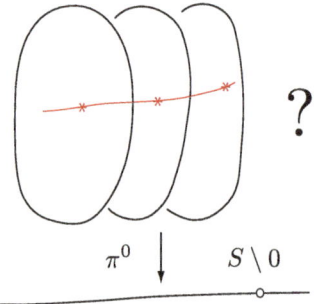

Figure 1.2: A one-parameter degeneration.

Assume that K_{X^0} is relatively ample over S^0. In this case, X^0 is equal to $\operatorname{Proj} R_{S^0}(X^0, K_{X^0})$. So the idea is to extend X^0 to some appropriate X and then take the relative canonical model. Since that model is unique and does not depend on the choice of an X, this will provide a unique extension in some sense. This quite simple idea has to be worked out more carefully.

Let us begin, more generally, with a pair $(X^0, B^0 = \sum b_i B_i^0)$ and with a morphism to S^0 such that the fibers $(X, B)_t$ are lc and have ample $\mathbb{R}$-Cartier $K_{X_t} + B_t$. After shrinking $(S, 0)$, one can assume that (X^0, B^0) is lc. Let $Y^0 \to X^0$ be a log resolution for the pair (X^0, B^0). After shrinking $(S, 0)$, one can assume that all the exceptional divisors E_j^0 are horizontal (the image is the whole S^0). At this point we have $X^0 = \mathrm{Proj}\, R_{S^0}(Y^0, K_{Y^0} + f_*^{-1} B^0 + \sum E_j^0)$.

Now apply a version of the Semistable Reduction Theorem to the normal crossing pair $(Y^0, \cup f_*^{-1} B_i^0 \cup E_j^0)$ which says that, after a base change $S' \to S$ (in order not to introduce horrible notation, we will skip the primes and denote the new curve again by S), there exists an extension $\pi : Y \to S$ such that Y is smooth, the central fiber Y_0 is a normal crossing divisor in which *every irreducible component has multiplicity 1*, and $\cup f_*^{-1} B_i \cup E_j \cup Y_0$ is a normal crossing divisor.

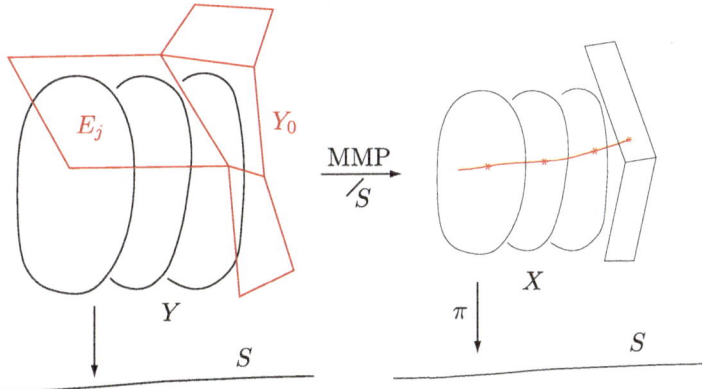

Figure 1.3: A one-parameter degeneration completed.

We now take the relative log canonical model

$$X := \mathrm{Proj}\, R_S\left(Y, K_Y + \sum b_i f_*^{-1} B_i + \sum E_j + Y_0\right).$$

Then, the horizontal divisors E_j collapse onto X, the pair $(X, \sum b_i B_i + X_0)$ has lc singularities, and $K_X + \sum b_i B_i + X_0$ is π-ample. Restricting to the central fiber, we see that the divisor $K_{X_0} + \sum b_i B_{i,0}$ is ample.

Because we insisted that Y has irreducible components of multiplicity 1, the log canonical model does not depend on the choice of Y by the uniqueness property of lc models, Lemma 1.2.4.

What singularities does the central fiber (X_0, B_0) have? We certainly cannot expect that X_0 will be normal: already a degeneration of curves is a stable curve which is typically not normal or irreducible. Whatever the singularities are, we should call them "semi-log canonical". The definition of [42] for surfaces used semi-log resolutions. In [2] an alternative definition was given, with the advantage that it is easier to work with it:

Definition 1.3.1. Let $(X, B = \sum b_i B_i)$ be a pair consisting of a (reduced) variety and a $\mathbb{Q}$-divisor or $\mathbb{R}$-divisor on it. Then it is called *semi-log canonical* (slc) if:

1. X satisfies Serre's condition S_2;

2. X has only double normal crossings in codimension 1, and the double locus has no irreducible components in common with B_i's;

3. the divisor $K_X + B$ is $\mathbb{R}$-Cartier (note that K_X is a well defined Weil divisor class because of the previous condition);

4. denoting by $\nu\colon X^\nu \to X$ the normalization, the pair $(X^\nu, \sum b_i \nu^{-1}(B_i) + D^\nu)$ is lc, where D^ν is the preimage of the double locus on X^ν.

Remark 1.3.2. S_2 is a natural generalization of "normal" to the case of varieties which may not be regular in codimension 1. A well-known theorem by Serre says that "normal" $= S_2 + R_1$. Removing R_1 leaves S_2.

One geometric consequence of S_2 is that an S_2 variety is "connected in codimension 1": one cannot disconnect it locally analytically by removing a subset of codimension bigger than or equal to 2. For example, a surface obtained from another surface by gluing together two points is not S_2. More generally, an S_2 variety can be uniquely reconstructed from any open subset U with $\mathrm{codim}(X \smallsetminus U) \geq 2$. In particular, a surface obtained by "pinching" a single point is not S_2 either.

Remark 1.3.3. One can define a divisorial sheaf $\mathcal{O}_X(K_X + B)$ on an slc variety as follows. The "bad" subset of X is the set where X has worse than double normal crossing singularities, plus $D \cap (\cup B_i)$. This set has codim $Z \geq 2$; let U be its complement. On U we have a well-defined dualizing sheaf ω_X, which is an invertible sheaf. We denote by K_U the linear equivalence class of divisors H on U such that $\omega_U \simeq \mathcal{O}_U(H)$. For any $d \in \mathbb{Z}$, one defines

$$\mathcal{O}_X\big(d(K_X + B)\big) := j_*\omega_X^{\otimes d}(dD).$$

Remark 1.3.4. For the normalization of a double normal crossing singularity, one has $\nu^*(K_X + B) = K_{X^\nu} + \nu^{-1}(B) + D^\nu$; so, condition (3) in Definition 1.3.1 is equivalent to saying that, for the log resolution $Y \to X^\nu \to X$ of the normalization of X, in the formula

$$K_Y \sim_{\mathbb{Q}} f^*(K_X + B) + \sum a_D D$$

all the discrepancies satisfy $a_D \geq -1$, just as in the definition of lc singularities.

Example 1.3.5. A curve pair $(X, \sum b_i B_i)$ with $b_i > 0$ is slc if and only if the singularities of X are at worst double normal crossings (locally analytically isomorphic to $xy = 0$), the divisors B_i do not contain the nodes, and for every non-nodal point $x \in X$ one has $\mathrm{mult}_x B = \sum b_i \mathrm{mult}_x B_i \leq 1$.

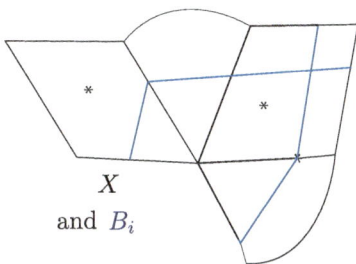

Figure 1.4: Semi log canonical singularities.

If B_i are points, then we are saying that they have to be nonsingular points of X, and they may coincide but when they do, the total weight has to be at most 1, just as in the definition of a weighted stable curve.

In our family, the pair $(X, \sum b_i B_i + X_0)$ is lc. The fact that the central fiber X_0 is S_2 was proved in [6]. By adjunction, this easily implies that the pair $(X_0, \sum b_i B_{i,0})$ is slc. This proves the existence part of the following theorem:

Theorem 1.3.6. Let $\pi^0 \colon (X^0, B^0) \to S^0$ be a family of irreducible slc pairs $(X, B)_t$ with ample $K_{X_t} + B_t$. Then, possibly after a finite base change $S' \to S$, this family can be uniquely extended to a complete family over S such that the central fiber (X_0, B_0) is slc with ample $K_{X_0} + B_0$.

The uniqueness part is proved as follows. We start with a completed family $\pi \colon (X, B) \to S$ extending π^0 such that the central fiber (X_0, B_0) is slc. From a highly nontrivial result from [34], namely the Inversion of Adjunction Theorem, it follows that the ambient family $(X, B + X_0)$ is lc. Let $f \colon Y \to X$ be a log resolution for the pair $(X, B + X_0)$. Denote by E_j the horizontal f-exceptional divisors; some irreducible components $(Y_0)_s$ of the central fiber may also be f-exceptional.

If $K_X + B$ is $\mathbb{Q}$-Cartier then X can be recovered as

$$X = \operatorname{Proj} R\Big(Y, K_Y + \sum f_*^{-1} B + \sum E_j + \sum (Y_0)_s\Big).$$

The last sum goes over all irreducible components of the central fiber Y_0, and we take them with coefficient 1 even if they have higher multiplicity in $Y_0 = \pi^*(0) = \sum m_s (Y_0)_s$. Now the uniqueness follows from Lemma 1.2.4.

Even if $K_X + B$ is only $\mathbb{R}$-Cartier, the uniqueness of the canonical model in the above sense is a bit harder to show, but still well known, see, e.g., [15].

See also [29].

1.4   Moduli of stable varieties

1.4.1   Definition of a stable pair

Theorem 1.3.6 is a sufficient motivation for the following definition:

Definition 1.4.1. A pair $(X, B = \sum b_i B_i)$ of a projective variety and an $\mathbb{R}$-divisor on it is called a *stable pair* if:

1. (singularities) (X, B) is slc (in particular, $K_X + B$ is an $\mathbb{R}$-Cartier divisor);
2. (numerical) $K_X + B$ is ample.

When $B = 0$, we simply talk about *stable varieties*.

1.4.2   Moduli functors in the case $B = 0$

We start with the case $B = 0$. In this case, there is a good introduction in [39], and all the necessary technical details have been filled in [38, 41].

The definition of a *family* of stable varieties of dimension bigger than or equal to 2 is nontrivial. Without taking care, it is easy to produce examples of flat families in which K_{X_t} is $\mathbb{Q}$-Cartier on each individual fiber but $K^2_{X_t}$ is not constant, see, e.g., [36]. To fix this, one has to make sure that NK_{X_t} comes from an invertible sheaf on the entire family X.

There are two basic functors, defined below. We use the terminology introduced in [31].

Definition 1.4.2. Let $d, N \in \mathbb{N}$ and $C \in \mathbb{R}$, $C > 0$. The *Viehweg moduli functor* $M_{N,C}$ is defined as follows: for any scheme S over the base field, $M_{N,C}(S)$ is the set of flat families $\pi : X \to S$ of dimension d such that

1. every geometric fiber X_t is a stable variety with $K^d_{X_t} = C$;
2. there exists an invertible sheaf $\mathcal{L}$ on X such that, for every geometric fiber X_t, one has
$$\mathcal{L}|_{X_t} \simeq \mathcal{O}_{X_t}(NK_{X_t}).$$

The *Kollár moduli functor* is defined as follows: $M_{K,C}(S)$ is the set of flat families $\pi : X \to S$ of dimension d such that

1. every geometric fiber X_t is a stable variety with $K^d_{X_t} = C$;
2. for *all* N the sheaves $F_N := j_* \mathcal{O}_U(NK_X)$ are flat over S and commute with arbitrary base changes $S' \to S$. Here, $U \subset X$ is the open subset on which π is relatively Gorenstein. Recall that, from the definition of slc, the restriction of $X \setminus U$ to each fiber has codimension bigger than or equal to 2.

Definition 1.4.3. The *moduli stacks* $\mathcal{M}_{N,C}$ and $\mathcal{M}_{K,C}$ are defined similarly: for any scheme S, the groupoid $\mathcal{M}_{N,C}(S)$, resp. $\mathcal{M}_{K,C}(S)$, is the category whose objects are the families as above and the arrows are isomorphisms over S.

1.4.3 Constructing the moduli spaces

There is a standard way to prove that the stacks $\mathcal{M}_{N,C}$ and $\mathcal{M}_{K,C}$ are algebraic and that the coarse moduli spaces for the moduli functors do exist. We briefly describe it. It uses MMP in a significant way (its full lc version) in dimension $\dim X + 1$ and boundedness, both of which are only available for $\dim X = 2$. Therefore, we restrict to this case and below we shall only consider stable surfaces.

1. The first step is the *Boundedness Theorem* (see [1]) which in fact is slightly stronger than what we formulate here; the full version allows for varying the coefficients b_i in a DCC set.

   Theorem 1.4.4 (Boundedness Theorem). *Fix $n \in \mathbb{N}$, $b_1, \ldots, b_n \in (0,1]$ and $C \in \mathbb{R}_{>0}$. Then the family of stable surfaces $(X, B = \sum_{i=1}^{n} b_i B_i)$ with $(K_X + B)^2 = C$ is bounded, i.e., there exist a scheme of finite type $\mathcal{S}$ and a family $(\mathcal{X}, \mathcal{B}_1, \ldots, \mathcal{B}_n) \to \mathcal{S}$ in which all such surface pairs appear as fibers.*

   In this section, we use this theorem for $n = 0$, i.e., $B = 0$. Below, we will use its full power. Importantly, the coefficients b_i are allowed to be non-rational, so that $K_X + B$ is only an ample $\mathbb{R}$-divisor.

   By boundedness, there exists a universal N such that the divisor $H = NK_X$ is invertible and very ample for all of our surfaces. Also, there exist only finitely many possibilities for $\chi(\mathcal{O}_X)$. Therefore, there are only finitely many possibilities for the Hilbert polynomial $p(d) = \chi(\mathcal{O}_X(dH)) = d^2 H^2/2 + dHK_X + \chi(\mathcal{O}_X)$.

   Let Hilb $= \bigcup \text{Hilb}_{p_i(d)}(\mathbb{P}^{N_i})$ be the finite union of the corresponding Hilbert schemes. The universal family $\mathcal{X} \to$ Hilb contains all of our surfaces, and also a bunch of other surfaces that we do not want. Now we need to weed them out.

2. A very delicate and technically difficult property is the *Local Closedness* of the moduli functor M, which says that for any family $\mathcal{X} \to \mathcal{S}$ of surfaces there exists a locally closed subscheme $T \to S$ with the following universal property: "*for any $S' \to S$, the family $\mathcal{X} \times_S S'$ obtained by base change is in $M(S)$ if and only if $S' \to S$ factors through $T \to S$*". For both Viehweg and Kollár moduli functors this property was established in full generality in [38].

   After applying this theorem to the family $\mathcal{X} \to$ Hilb, we now have a family $\mathcal{X}_T \to T$ which contains all of our surfaces and only them.

   Next, we have to take the quotient of T by an appropriate equivalence relation R so that T/R is our coarse moduli space, and the stack $[T/R]$ is our algebraic moduli stack. There are convenient and powerful theorems for that, see [35, 37], but they only work for a proper equivalence relation, so one first has to establish a couple more properties.

3. *Finite Automorphisms.* This property says that for any stable pair the automorphism group $\text{Aut}(X, B)$ is finite. This easily follows from an old and

general result of Iitaka saying that for any smooth projective variety Y and a normal crossing divisor D on it such that $K_Y + D$ is big, the group $\mathrm{Aut}(Y, D)$ is finite. We apply it to the resolution of singularities of each irreducible component of the normalization of X, and D is the support of the preimages of B_i and the exceptional divisors. Also, (3) essentially follows from the following condition (4).

4. *Properness of the functor.* This is the property with which we started the discussion: for any family $\pi^0 \colon X^0 \to S^0$ over a punctured curve $S^0 = S \smallsetminus \{0\}$, after possibly a finite base change $S' \to S$, there exists a unique way to extend it to a complete family $\pi \colon X \to S$.

   However, we only established it in the case when the variety X^0 is normal, i.e., the generic fiber is irreducible. One needs it in the general case.

   Intuitively, the general case is reduced to the irreducible case by normalizing X^0, finding the limits for each irreducible component (X_k^0, D_k^0), where D_k^0 is the double locus, and then gluing the limits into a total family. The gluing should work nicely by the uniqueness property of the log canonical model and compatibility with adjunction to D_k.

   In reality, the process is quite delicate. One has to realize the crucial point that, for the gluing to work nicely, the so-called *differents* (whose definition we skip) on both sides of the double locus have to match. The good news are that, for surfaces, Kollár proved all the necessary results, see [40, 41], so this property has been established.

5. At this point, everything is set to take the quotient. First, one has to make sure that in Step 1 the multiple $H = NK_X$ is taken large enough so that not only H is very ample, but also has no higher cohomology: for all of our pairs $H^p(X, \mathcal{O}(H)) = 0$ for $p > 0$. Again, by boundedness, such a multiple exists.

   So at this time we have finitely many locally closed subschemes T of Hilbert schemes $\mathrm{Hilb}_{p(d)}(\mathbb{P}^N)$, and all of our surfaces appear in the universal families $\mathcal{X}_T \subset \mathbb{P}^N \times T$. The group $\mathrm{PGL}(N+1)$ acts on T and sends surfaces to isomorphic surfaces.

   Moreover, for any family $\pi \colon X_S \to S$ in $M(S)$ (where $M = M_{N,C}$ or $M_{K,C}$) the push forward $\pi_* \mathcal{O}_{X_S}(H) \to S$ is a locally free sheaf on X of rank $N+1$. This follows from the condition $H^p(X, \mathcal{O}(H)) = 0$ for $p > 0$ by the Cohomology and Base Change Theorem.

   On an open cover $S = \bigcup U_\alpha$ the sheaf becomes free: $(\pi|_{U_\alpha})_* \mathcal{O}_{X_{U_\alpha}}(H) \simeq \mathcal{O}_{U_\alpha}^{\oplus N+1}$. Choosing a basis, i.e., a concrete isomorphism $(\pi|_{U_\alpha})_* \mathcal{O}_{X_{U_\alpha}} \to \mathcal{O}_{U_\alpha}^{\oplus N+1}$ gives a map $S \to T$ so that locally on the base S, X_{U_α} is the pullback of the universal family $X_T \to T$. Any other isomorphism differs from this by an element in $\mathrm{PGL}(U_\alpha)$.

   This implies that the moduli stack $\mathcal{M}$ is the quotient $[T / \mathrm{PGL}(N+1)]$. The group action is proper by the properties 4. and 5. Now, by [35], the quotient is an algebraic stack with a coarse moduli space which is a proper algebraic space.

6. Finally, the coarse moduli space M is not just a proper algebraic space, but it is a projective scheme by Kollár [36]. The result of [36] is only for surfaces. It was extended to higher dimensions and to pairs $(X, \sum b_i B_i)$ by Fujino [20].

1.5 Moduli of stable pairs (X, B) with $B \neq 0$

1.5.1 A tricky problem

Practically all of the steps of the previous section go over verbatim to the case of surface pairs $(X, B = \sum b_i B_i)$. However, there is the following problem that was identified by Hacking.

For simplicity, let us assume that $B = b_1 B_1$ is irreducible and that $b_1 \in \mathbb{Q}$. Let $\pi^0 \colon X^0 \to S^0$ be a one-parameter degeneration which we completed to a family $\pi \colon X \to S$. By construction, the divisor $K_X + B$ is $\mathbb{Q}$-Cartier. It is however possible that the divisors K_X and B are not $\mathbb{Q}$-Cartier individually. In this case, in the central fiber, the closed subscheme $B_1 \cap X_0 \subset X_0$ may have an embedded prime. In other words, it will not be a divisor, but only a closed subscheme. Hassett came up with a concrete example with coefficient $b_1 = 1/2$, which is reproduced in [6].

Indeed, by the properness of Hilbert schemes, we know that a limit of subvarieties in a one-parameter family exists, but in general it is only a subscheme, not a subvariety. It may have nilpotents, and may have embedded components. This is what happens here.

On the face of it, this problem means that perhaps a whole approach has to be rethought from the ground up. Perhaps one has to set up the theory of stable pairs as pairs $(X, \sum b_i B_i)$ where $B_i \subset X$ are subschemes.

This has many unwanted consequences. For example, what if some of the embedded points in B_i "wander away" from the divisorial part after a deformation? Do we have to track such "free floating points" now? That makes the moduli functor a lot bigger that desired.

Below we list several known solutions to this problem. We also provide a new solution for very generic coefficients in Subsection 1.5.3.

1.5.2 Large coefficients

In [6] it was shown that the problem does not appear for the components B_i with $b_i = 1$. Subsequently, Kollár [41] improved this significantly by showing that for any component with coefficient $b_i > 1/2$ the divisor B_i does not acquire embedded primes and stays a divisor. So the boundary $b_i = 1/2$ in Hassett's example is the best possible. Therefor, for as long as all $b_i > 1/2$, everything works and we have a moduli space.

1.5.3   Very generic coefficients

We start with the following elementary lemma.

Lemma 1.5.1. *Let $b_0 = 1, b_1, \ldots, b_n$ be real numbers which are linearly independent over $\mathbb{Q}$, and suppose that the divisor $\sum_{i=0}^n b_i B_i$ is $\mathbb{R}$-Cartier. Then each of the divisors B_i is $\mathbb{Q}$-Cartier.*

Proof. Indeed, extend $b_0, \ldots, b_n$ to a basis $\{b_i, \ i \in I\}$ of $\mathbb{R}$ as a $\mathbb{Q}$-vector space (of course, the index set I is uncountable). The divisor $\sum_{i=0}^n b_i B_i$ being $\mathbb{R}$-Cartier means that

$$\sum_{i=0}^n b_i B_i = \sum d_k D_k$$

for some real numbers d_k and $\mathbb{Z}$-divisors D_k. Expand each of the coefficients d_k in the above basis: $d_k = \sum_i d_{k,i} b_i$, a finite sum with $d_{k,i} \in \mathbb{Q}$. Then the above equality is equivalent to

$$B_i = \sum_k d_{k,i} D_k \quad \text{for } i = 0, \ldots, n \quad \text{and} \quad \sum_k d_{k,i} D_k = 0 \quad \text{for } i \neq 0, \ldots, n.$$

Hence, the divisors B_i are $\mathbb{Q}$-Cartier.   □

We can apply the above lemma to the divisor $K_X + B$, $B = \sum_{i=1}^n b_i B_i$. Then in the completed one-parameter family the divisors K_X and B_i stay $\mathbb{Q}$-Cartier, and the problem disappears.

Remark 1.5.2. Here is a way to understand this trick. If we start with a $\mathbb{Q}$-factorial family $X^0 \to S^0$, then a minimal model $X_{\min} \to S$ of a semistable model $Y \to S$ will still be $\mathbb{Q}$-factorial, and the divisors B_i will still be $\mathbb{Q}$-Cartier.

It is on the last step, going from the minimal to the canonical model, that some curves C in the central fiber may get contracted such that $B_i C \neq 0$. This forces the divisors B_i on the log canonical model to not be $\mathbb{Q}$-Cartier.

The curves that get contracted satisfy the equation $(K_X + \sum b_i B_i)C = 0$. This gives finitely many linear equations, defining finitely many hyperplanes. For a generic (b_i) lying outside of the hyperplanes, the curves are not contracted, and the divisors B_i should stay $\mathbb{Q}$-Cartier.

So this should be a general picture: there should exist a locally finite chamber decomposition and, for any (b_i) in a maximal-dimensional chamber, the divisors B_i should stay $\mathbb{Q}$-Cartier. To make this into a proof, however, one has to consider all one-parameter degenerations for all (b_i), etc. Choosing b_i so generic that they are linearly independent over $\mathbb{Q}$ provides a quick solution.

At this point we solved the problem with embedded primes. However, in both Viehweg's and Kollár's moduli functors, it is necessary that some multiple $N(K_X + B)$ is a Cartier $\mathbb{Z}$-*divisor*. In Viehweg's functor it is explicit, in Kollár's functor it is needed to make the number of conditions to check finite (since we are applying the local closedness to each multiple $N \in \mathbb{N}$). Here is how to fix this small obstacle.

Fix n and $\mathbb{Q}$-linearly independent $(1, b_1, \ldots, b_n)$. Next, instead of fixing just one number $(K_X + B)^2$, fix a vector $\boldsymbol{C}$ of all possible top intersection products $K_X^{k_0} B_1^{k_1} \cdots B_n^{k_n}$ with $\sum k_i = 2$. Since all the divisors are $\mathbb{Q}$-Cartier, these are well-defined rational numbers.

The Boundedness Theorem 1.4.4 applies to $\mathbb{R}$-divisors, so it says that the family of pairs $(X, B = \sum b_i B_i)$ with $\mathbb{Q}$-Cartier divisors B_i and ample $\mathbb{R}$-Cartier divisors B_i is bounded. Therefore, there exist some nearby rational numbers b_i' for which the divisors $K_X + B'$, $B' = \sum b_i' B_i$ are ample for all of our pairs. We can easily compute the new vector $\boldsymbol{C}'$ for these modified coefficients.

The family of *all* pairs with vector $\boldsymbol{C}'$ for which $K_X + \sum b_i' B_i$ is ample is bounded by the same Theorem 1.4.4. The subset of the base for which the fibers have no embedded primes is open. To this subset we can now apply the Local Closedness Theorem to the integral multiples of $K_X + B'$. Then in the resulting family $X_T \to T$, we pick an open subset of T parameterizing the pairs with ample $K_X + \sum b_i B_i$.

The rest of the proof proceeds as before. Most crucially, the functor is proper.

1.5.4 Very generic coefficients and $K_X \sim_{\mathbb{Q}} 0$

Let us note one special case from the previous subsection. The moduli functor for the stable pairs $(X, B = \sum_{i=1}^{n} b_i B_i)$ satisfying the following four conditions has a coarse moduli space which is a projective scheme: (1) the coefficients $1, b_1, \ldots, b_n$ are linearly independent over $\mathbb{Q}$; (2) K_X and B_i are $\mathbb{Q}$-Cartier; (3) some multiple $N K_X$ is linearly equivalent to zero; and (4) $(K_X + B)^2$ is constant.

To the previously established result, we need to add the following well-known statement which allows to carve out the subfamily where the sheaf $\mathcal{O}_X(N K_X)$ is zero on the fibers.

Lemma 1.5.3. *Let $\pi \colon X \to S$ be a flat projective family with geometrically reduced connected fibers and L be an invertible sheaf on X. Then there exists a closed subscheme $T \subset S$ satisfying the following universal condition: for any base change $S' \to S$, on the family $X' = X \times_S S' \to S'$ the invertible sheaf $L' = g^* L$ is the pullback of an invertible sheaf from the base if and only if the morphism $S' \to S$ factors through T.*

Proof. See, e.g., [61, Lemma 1.19]. $\qquad\square$

In some situations, even stronger results are available. For example, one has the following:

Theorem 1.5.4. *Fix an integer C. Let $\pi^0 \colon (X^0, \epsilon H^0) \to S^0$ be a degenerating one-parameter family of stable pairs in which the fibers are either abelian surfaces or K3 surfaces, and H is an effective ample Cartier divisor with $H^2 = C$. Then, perhaps after a finite base change $S' \to S$, there exists an extension $\pi \colon X \to S$ in* *which the central fiber satisfies $K_{X_0} \sim 0$, H_0 is Cartier, and the pair $(X_0, \epsilon H_0)$ has slc singularities for all $\epsilon < \epsilon_0(C)$.*

Proof. For abelian varieties of any dimension, the proof is contained in [4]. For K3 surfaces, a sketch of the proof was given in [43] which is somewhat incomplete, but can be fixed. To complete it, one has to observe that Shepherd–Barron's operations for a degeneration of K3 surfaces preserve the condition for H to be Cartier. ☐

Corollary 1.5.5. *For any $d \in 2\mathbb{N}$ there exists a small irrational ϵ such that the moduli space P_d of stable K3 surface pairs $(X, \epsilon H)$ such that $H^2 = d$ is an open subset of a coarse moduli space $\overline{P}_d$ of stable slc pairs $(X, \epsilon H)$. Furthermore,*

1. *there exists $N \in \mathbb{N}$ such that for all stable pairs parameterized by $\overline{P}_d$ one has $N K_X \sim 0$;*

2. *for any family in the closure of P_d in $\overline{P}_d$, one has $K_X \sim 0$ and H is Cartier.*

Note that, generally, $\overline{P}_d$ may have several irreducible components. The theorem above guarantees $K_X \sim 0$ and H Cartier only for the pairs in the main irreducible component of $\overline{P}_d$, for the "smoothable" pairs.

1.5.5   Pairs $(X, \sum b_i B_i)$ with branchdivisors $B_i \to X$

Another solution to the problem of subschemes $B_i \subset X$ with embedded primes is to replace them with branchvarieties $B_i \to X$ introduced in [12]. A branchvariety over a projective variety X is a *reduced* variety B_i together with a *finite* morphism $B_i \to X$. Thus, we trade an embedded possibly nonreduced subscheme for a reduced variety, but only with a finite morphism. [12] proves that the moduli of branchvarieties are proper, so any one-parameter degeneration has a unique limit.

Again, introducing brachdivisors leads to more pairs than perhaps desirable. For example, for the coefficient $b_i = 1/2$, instead of just considering a divisor $b_i B_i$ in which B_i perhaps has a component of multiplicity 2, we must consider all double covers $B_i \to \mathrm{im}\, B_i \subset X$, and there are lots of them.

1.5.6   Replace a divisor by a sheaf homomorphism

Kollár suggested the following solution. Let $B = \sum b_i B_i$ be an effective $\mathbb{Q}$-divisor such that NB is a $\mathbb{Z}$-divisor and $\mathcal{O}_X(N(K_X + B))$ is an invertible sheaf. Then we can encode B by the homomorphism $\varphi \colon \omega_X^{\otimes N} \to L$. Symbolically, "$B = (K_X + B) - K_X$".

Thus, a family of pairs $(\mathcal{X}, \mathcal{B}) \to \mathcal{S}$ can be encoded by an invertible sheaf L on $\mathcal{X}$ and a homomorphism of sheaves $\omega_{\mathcal{X}/\mathcal{S}}^{\otimes N} \to L$. The sheaf $\omega_{\mathcal{X}/\mathcal{S}}^{\otimes N}$ may be very nasty, have torsion and cotorsion, etc. But, its formation commutes with arbitrary base changes. Of course, the same is true for L because a pullback of an invertible sheaf is invertible.

Thus, one obtains a well-defined functor with nice properties. The fact that it has the Local Closedness property follows from [38]. The rest of the construction of the moduli space should proceed as before.

1.6 Moduli of stable varieties and pairs: known cases

We list the known cases where the moduli spaces of higher-dimensional stable pairs are known to exist, and perhaps something more than just an existence theorem is available.

1.6.1 Surfaces and some surface pairs (X, B)

Let us state clearly which results for surfaces are well established, with a complete proof available.

Recall that for stable varieties X, the Viehweg moduli functor $M_{N,C}$ and the Kollár moduli functor $M_{K,C}$ were defined in 1.4.2. For the pairs $(X, B = \sum b_i B_i)$ with fixed rational b_i, the functors are defined in the same way, with the multiples NK_X replaced by the multiples $N(K_X + B)$ for which all Nb_i are integral. Finally, for the real numbers b_i, the coefficients are replaced by nearby rational numbers b_i', as in Subsection 1.5.3.

Theorem 1.6.1. *For fixed n, $(b_1, \dots, b_n)$, and C, for stable surface pairs $(X, B = \sum_{i=1}^{n} b_i B_i)$, both the Viehweg moduli functor with appropriate $N(n, b_i, C)$ and appropriate (b_i') as at the end of Subsection 1.5.3, and the Kollár moduli functor with appropriate (b_i') are coarsely represented by projective schemes in all of the following cases:*

1. *$B = 0$;*

2. *large coefficients: all $b_i > 1/2$;*

3. *very generic coefficients: $(1, b_1, \dots, b_n)$ are linearly independent over $\mathbb{Q}$;*

4. *large and very generic coefficients: some b_i are rational and $b_i > 1/2$, and the others are real and linearly independent over $\mathbb{Q}$;*

5. *very generic coefficients and $K_X \sim_{\mathbb{Q}} 0$, i.e., some positive multiple $NK_X \sim 0$.*

1.6.2 Products of curves and similar surfaces

By [58, 60], the stable limit of a family of surfaces which are products of smooth curves $X_t = C_t \times C_t'$ or symmetric powers $X_t = (C_t \times C_t)/\mathbb{Z}_2$ is again a product or a symmetric power of *stable curves* C_0, C_0'. Thus, the compactification of this component in the moduli space of surfaces of general type is $\overline{M}_g \times \overline{M}_{g'}$, resp. $\overline{M}_g$.

This was generalized to some surfaces which are finite quotients of products of curves by more interesting automorphism groups in [44, 59].

1.6.3   Planar curve pairs

Hacking [26] considered compactifications of moduli spaces of pairs $(\mathbb{P}^2, (\frac{3}{d}+\epsilon)C)$, where C is a curve of degree d and $0 < \epsilon \ll 1$. (Note that $K_{\mathbb{P}^2} + (\frac{3}{d}+\epsilon)C \sim d\epsilon H$ is very small.)

He proved that, when d is not divisible by 3, the compactified moduli stack is smooth. He also provided a rough classification of the degenerate stable pairs.

1.6.4   Del Pezzo surface pairs

Hacking and Tevelev worked out in [28] several cases of compactifications of surface pairs $(X, \sum B_i)$ where X is a del Pezzo surface and B_i are the lines, i.e., (-1)-curves.

1.6.5   Special surfaces of general type

As an application of the theory of weighted stable hyperplane arrangements, Alexeev in [7] and Alexeev and Pardini in [13] explicitly computed degenerations for several types of surfaces of general type, including some numerical Campedelli surfaces and Burniat surfaces.

1.6.6   Stable toric varieties

Toric varieties give stable pairs in a very simple way. If X is a toric variety and Δ is the union of boundary divisors, then the pair (X, Δ) is lc and $K_X + \Delta \sim 0$. A pair $(X, \Delta + \epsilon B)$ for $0 < \epsilon \ll 1$ is a stable pair if and only if B does not contain any torus strata and B is ample. Thus, this case corresponds to the coefficients $b_1 = 1$, $b_2 = \epsilon$.

The moduli of stable toric pairs provides a compactification. It was constructed in [4] and [9]. It was further extended to spherical varieties in [8, 9].

1.6.7   Abelian varieties

If X is an abelian variety or an abelian torsor (a principally homogeneous space) over an abelian variety and B is an ample divisor ("theta divisor"), then $(X, \epsilon B)$ is a stable pair. The compactification using stable semiabelic varieties was constructed in [4]. Formally, the theory is the infinite-periodic analogue of the theory of stable toric varieties.

1.6.8   Weighted stable hyperplane arrangements

The moduli of weighted stable hyperplane arrangements [7] provides the compactification for the moduli space of lc hyperplane arrangements $(\mathbb{P}^{r-1}, \sum b_i B_i)$ with $\sum b_i > r$. This is the major topic of these lectures. The case of the weights $b_i = 1$ is contained in [27].

Chapter 2

Stable Toric Varieties

For a detailed introduction to the theory of toric varieties, one should consult the usual sources [21, 48]. Our introduction is very brief and serves mainly to set up the notation and clarify the definitions (for example, our toric varieties are normal and do not have the "origin" fixed). The theory of stable toric varieties reviewed below is contained in [4, 9].

By k we denote the base field, which is assumed to be algebraically closed.

2.1 Projective toric varieties and polytopes

2.1.1 Toric varieties and torus embeddings

The *multiplicative group variety* $\mathbb{G}_m$ is the group variety $\mathrm{Spec}\, k[t, 1/t] = \mathbb{A}^1 \setminus \{0\}$. It comes with the structure morphisms $\mathrm{mult}\colon \mathbb{G}_m \times \mathbb{G}_m \to \mathbb{G}_m$, $\mathrm{inverse}\colon \mathbb{G}_m \to \mathbb{G}_m$, and $\mathrm{unit}\colon \mathrm{Spec}\, k \to \mathbb{G}_m$, satisfying the group axioms. The reason not to write simply k^* is that k^* is a set but $\mathbb{G}_m$ is an algebraic variety. The set of k-points is $\mathbb{G}_m(k) = k^*$.

The *multiplicative torus* T of dimension r is $\mathbb{G}_m^r = \mathbb{G}_m \times \cdots \times \mathbb{G}_m$. It comes with two standard lattices commonly denoted M and N, $N = \mathrm{Hom}(M, \mathbb{Z})$:

1. The lattice $N = \mathrm{Hom}(\mathbb{G}_m, T) \simeq \mathbb{Z}^r$ of *one-parameter subgroups*. An arbitrary toric variety is described by a fan (a collection of finitely generated strictly convex cones) in $N_{\mathbb{R}} = N \otimes \mathbb{R}$. The pictures recorded in this space are "inverted". A cone τ of dimension d corresponds to a T-orbit O_τ of dimension $r - d$, and the order is reversed: τ_1 is a face of τ_2, denoted $\tau_1 \prec \tau_2$, if $\overline{O}_{\tau_1} \supset O_{\tau_2}$.

2. The lattice $M = \mathrm{Hom}(T, \mathbb{G}_m) \simeq \mathbb{Z}^r$ of *characters*, or *monomials*. This space is responsible for a "direct picture". A d-dimensional polytope in $M_{\mathbb{R}} = M \otimes \mathbb{R}$ corresponds to a d-dimensional projective toric variety, and the inclusions go the same way.

We will work exclusively with the M-lattice, which in fact is much easier.

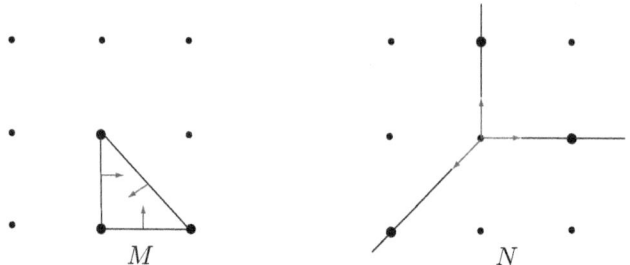

Figure 2.1: A polytope in M and its normal fan in N.

Definition 2.1.1. A *toric variety* is a normal variety with a T-action which has a dense T-orbit O. A *torus embedding* is a normal variety with a T-action and a fixed embedding $T \subset X$ which is a dense T-orbit.

Thus, the main difference between a toric variety and a torus embedding is that the latter comes with a special point $1 \in T \subset X$, while in the former the "origin" is not chosen.

If p is any point in the dense orbit O and T_p is its stabilizer, then the orbit O is a torsor (principal homogeneous space) over a torus $T'' = T/T_p$ of dimension at most r. We allow the stabilizer T_p to be nontrivial but we assume that it is connected. In characteristic $p > 0$ we assume additionally that it is reduced. As any algebraic subgroup of a torus, T_p is the product of a torus T' and several copies of groups of roots of unity μ_{n_i} (which in characteristic p may be connected and non-reduced if n_i is a power of p). So we are saying that $T_p = T'$ and there is no finite part.

A torus embedding together with a T-action has no isomorphisms. A toric variety together with a T-action still has an automorphism group equal to T''.

2.1.2 Polarized toric varieties vs polytopes

Definition 2.1.2. A *polarized toric variety* is a pair (X, L) of a projective toric variety X and an ample line bundle L on it.

Every line bundle on a toric variety is linearizable, and two linearizations differ by an element in $\mathrm{Hom}(T, \mathbb{G}_m) = M$. A *linearization* of L is a lift of the action $T \curvearrowright X$ to the $\mathbb{A}^1$-bundle $\mathbb{L} \to X$ corresponding to L. When L is ample, it is also the same as an action of T on the ring $R(X, L) = \bigoplus_{d=0}^{\infty} H^0(X, L^d)$ which induces the original T-action on $X = \mathrm{Proj}\, R(X, L)$.

The main connection that we need between algebraic geometry and combinatorics is expressed in the following result.

Theorem 2.1.3. *There is a 1-to-1 correspondence between (isomorphisms classes of) polarized toric varieties (X, L) with linearized L, and polytopes P with vertices in the lattice M, under which $\dim X = \dim P$.*

Furthermore, a polytope P_1 is a face of a polytope P_2 if and only if X_1 is a T-invariant subvariety of X_2 and $L_1 \simeq L_2|_{X_1}$ as T-linearized line bundles.

Translating a polytope by an element of M corresponds to another choice of the linearization.

Figure 2.2 illustrates this correspondence.

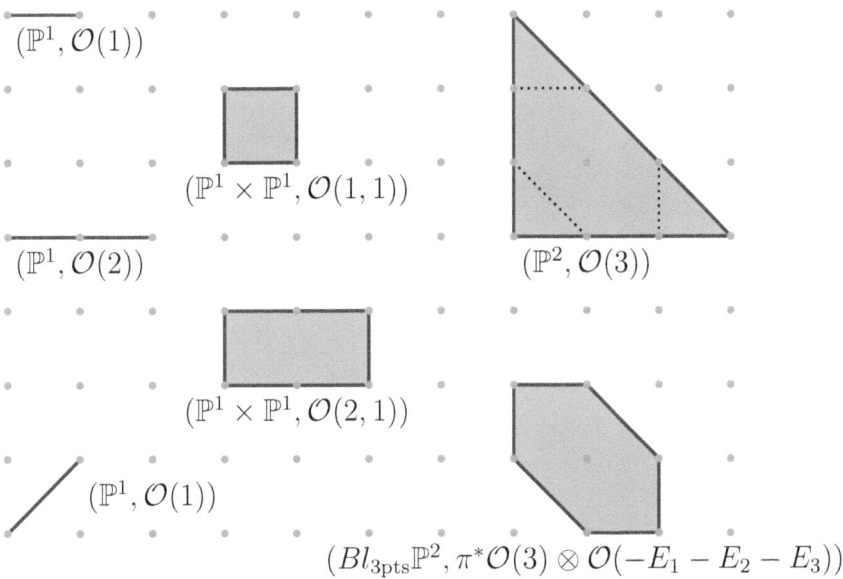

$(\mathbb{P}^1, \mathcal{O}(1))$

$(\mathbb{P}^1 \times \mathbb{P}^1, \mathcal{O}(1,1))$

$(\mathbb{P}^1, \mathcal{O}(2))$

$(\mathbb{P}^2, \mathcal{O}(3))$

$(\mathbb{P}^1 \times \mathbb{P}^1, \mathcal{O}(2,1))$

$(\mathbb{P}^1, \mathcal{O}(1))$

$(Bl_{3\text{pts}}\mathbb{P}^2, \pi^*\mathcal{O}(3) \otimes \mathcal{O}(-E_1 - E_2 - E_3))$

Figure 2.2: Polarized toric varieties $\leftrightarrows$ lattice polytopes.

The correspondence proceeds as follows. If (X, L) is a polarized toric variety and L is T-linearized, then T acts on $H^0(X, L)$. An algebraic action of a torus on a vector space V is diagonalizable and decomposes V into a direct sum $\bigoplus_{m \in M} V_m$ over the character group, so that for $v \in V_m$ the action is

$$\left(\lambda_1, \ldots, \lambda_r\right) \cdot v = \prod_{i=1}^{r} \lambda_i^{a_i} \cdot v,$$

where $m = (a_1, \ldots, a_r) \in M$. The characters m are also called *weights*. It is known that, for the action $T \curvearrowright H^0(X, L)$, the weights m with $V_m \neq 0$ are in bijection with the integral points of a lattice polytope P, and for each of these weights $\dim V_m = 1$. Thus, the polytope P is the convex hull of the weights m such that $H^0(X, L)_m \neq 0$.

In the opposite direction, start with a polytope $P \subset M_{\mathbb{R}}$. Let $\text{Cone}(1, P)$ be the cone in $\mathbb{R}^{1+r}$ over the polytope $(1, P)$. We call the extra dimension the degree, so we put P in degree 1. Let S be the semigroup of integral points $\mathbb{Z}^{1+r} \cap$

Cone$(1, P)$. It is graded by the degree. The semigroup algebra $k[S]$ is a graded algebra. Then, $X = \operatorname{Proj} k[S]$ and $L = \mathcal{O}_{\operatorname{Proj} k[S]}(1)$.

The elements $s = (d, m) \in S$ are the monomials x^s in this algebra, $\deg s = d$. Relations between the vectors in S give relations in $k[S]$. Choosing generators of S and figuring out relations between them gives concrete coordinates and homogeneous equations for X.

Example 2.1.4. Let P be the triangle with vertices $(0,0)$, $(0,1)$, $(1,0)$. Denote $u = x^{(0,0)}$, $v = x^{(1,0)}$, $w = x^{(0,1)}$. Then u, v, w generate $k[S]$ and there are no relations, so X is $\mathbb{P}^2$ with homogeneous coordinates u, v, w.

Let P be the square with vertices $(0,0)$, $(0,1)$, $(1,0)$, $(1,1)$. Denote $u = x^{(0,0)}$, $v = x^{(1,0)}$, $w = x^{(0,1)}$, $t = x^{(1,1)}$. Then u, v, w, t generate $k[S]$ and there is a single relation $(0,0) + (1,1) = (1,0) + (0,1)$. So X is a subvariety of $\mathbb{P}^3$ defined by the homogeneous equation $ut = vw$. Of course, $X \simeq \mathbb{P}^1 \times \mathbb{P}^1$.

If F is a polytope, then F^0 denotes its relative interior, i.e., F minus the proper faces. F^0 is a locally closed subset of $\mathbb{R}^r$.

Lemma 2.1.5. *There is a bijection between the T-orbits of X and locally closed faces F^0 of the polytope P, $O_F \leftrightarrows F$, $F \prec P$. Moreover, this bijection is dimension- and order-preserving:*

1. $\dim F = \dim O_F$, *and*
2. $F_1 \prec F_2$ *(i.e., $F_1^0 \subset \overline{F}_2^0$) if and only if $O_{F_1} \subset \overline{O}_{F_2}$.*

    Note that $P = \bigsqcup F_i^0$ and $X = \bigsqcup O_{F_i^0}$.

2.2   Stable toric varieties and tilings

A stable toric variety is a seminormal union of toric varieties, glued along T-invariant subvarieties. Combinatorially, it corresponds to a union of polytopes glued along faces.

Definition 2.2.1. A variety X is said to be *seminormal* if any finite morphism $f: X' \to X$ which is a bijection is in fact an isomorphism.

An example of a *non*-seminormal variety is the cuspidal curve $y^2 = x^3$: the normalization is a bijection, but not an isomorphism. This is a good way to think about seminormal varieties: they are varieties without "cusps". The main statement about seminormal singularities is the following one.

Lemma 2.2.2. *Any variety has a unique seminormalization $\pi^{\mathrm{sn}}: X^{\mathrm{sn}} \to X$, a proper bijective morphism with seminormal X^{sn} which has a universality property: any morphism $Y \to X$ from a seminormal variety factors uniquely through π^{sn}.*

A curve is seminormal if and only if it is locally analytically isomorphic to a union of n coordinate axes in $\mathbb{A}^n$ for some n. For such a curve, $n = \dim T_{X,x}$ at the singular point. In particular, a planar curve is seminormal if it has at worst nodes as singularities.

Definition 2.2.3. A polarized stable toric variety is a pair (X, L) of a projective variety with a linearized ample line bundle such that

1. X is seminormal, and
2. the irreducible components $(X_i, L_i = L|_{X_i})$ are polarized toric varieties.

(A "variety" for us need not be irreducible, but it has to be reduced; also, recall that our toric varieties are normal by definition.)

Thus, a stable toric variety is glued from ordinary toric varieties in a generic way, without introducing "cusps".

For every irreducible component (X_i, L_i) we have a lattice polytope. An intersection $X_i \cap X_j$ has to be T-invariant, so it is a closed union of orbits of both X_i and X_j. On the combinatorial side, this gives a closed union of faces of both P_i and P_j.

Definition 2.2.4. The *topological type* of a stable toric variety is the topological space $|\Delta| = \bigcup P_i$, a union of polytopes glued in the same way as $X = \bigcup X_i$, together with the finite map $\rho:|\Delta| \to M_{\mathbb{R}}$ such that $\rho|_{P_i}: P_i \to M_{\mathbb{R}}$ are the embeddings of lattice polytopes corresponding to (X_i, L_i).

The easiest complex Δ is a tiling of a bigger polytope P by smaller polytopes $P = \bigcup P_i$. An example is given in Figure 2.3. In these lectures, we will only work with stable toric varieties of this form. But in principle, the images of the polytopes in $M_{\mathbb{R}}$ are allowed to intersect or cover each other, so ρ is not necessarily an inclusion. For example, we can take two copies of the same square and glue them along the boundary; in this case ρ would be generically 2-to-1.

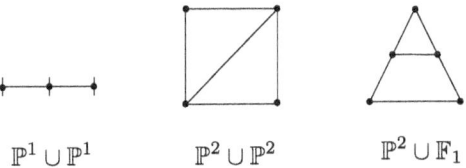

$$\mathbb{P}^1 \cup \mathbb{P}^1 \qquad \mathbb{P}^2 \cup \mathbb{P}^2 \qquad \mathbb{P}^2 \cup \mathbb{F}_1$$

Figure 2.3: Some simple stable toric varieties.

For each tiling Δ (or a more general complex of lattice polytopes), there is generally not a single variety, but a *family* of polarized stable toric varieties. That is because there may be many ways to glue the individual toric varieties.

The gluing can be understood as follows. Choose an "origin" in every irreducible component, thus fixing on each X_i the structure of a torus embedding. Then on every component of the intersections $X_i \cap X_j$ one gets two "origins", coming from X_i and from X_j. They differ by an element $t_{ij} \in T_{ij}$ in a corresponding torus. The collection (t_{ij}) has to satisfy the 1-cocycle condition $t_{ij}t_{jk}t_{ki} = 1$ on $X_i \cap X_j \cap X_k$. On the other hand, the "origins" in the varieties X_i can be chosen arbitrarily, up to an action of the tori T_i. Thus, the collection (t_{ij}) is defined only up to a 1-coboundary $(t_i t_j^{-1})$. Putting this together shows that the possible glued stable toric varieties X are in bijection with a certain 1-cohomology group $H^1(\Delta, \underline{T})$. Figure 2.4 gives an example where this group is one-dimensional.

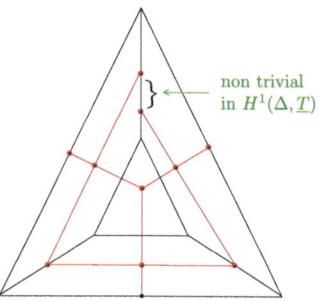

Figure 2.4: Complex Δ with a one-dimensional family of STVs.

Repeating the same argument for the *polarized* stable toric varieties we get that (X, L) are in bijection with the 1-cohomology group $H^1(\Delta, \mathbb{T})$, where $\underline{T}$ and $\mathbb{T}$ are constructible sheaves on $|\Delta|$ related by an exact sequence

$$1 \longrightarrow \underline{\mathbb{G}}_m \longrightarrow \mathbb{T} \longrightarrow \underline{T} \longrightarrow 1,$$

and $\underline{\mathbb{G}}_m$ is a constant sheaf on $|\Delta|$. In particular, if $|\Delta|$ is simply connected then $H^p(\Delta, \underline{\mathbb{G}}_m) = 1$ for $p > 0$ and $H^1(\Delta, \mathbb{T}) = H^1(\Delta, \underline{T})$.

2.3   Linear systems on toric and stable toric varieties

2.3.1   Linear systems on toric varieties

Let (X, L) be a polarized toric variety. Another basic fact is that the linear system $|L|$ is base-point free and defines a finite morphism $\varphi_L \colon X \to \mathbb{P}^N$, $N = h^0(X, L) - 1$, which however need not be a closed embedding or even generically 1-to-1.

Lemma 2.3.1. *Let m be an integral point in the lattice polytope associated to (X, L), and $e_m \in H^0(X, L)$ a corresponding section. Let F be the minimal face of P containing m, so that $m \in F^0$. Then*

1. *the open subset $U_m = \{e_m \neq 0\}$ is $U_m = \bigcup_{m \in F_i} O_{F_i^0}$, and*
2. *the closed subset $Z_m = (e_m)$ is $Z_m = \bigcup_{m \notin F_i} O_{F_i^0}$.*

This lemma is illustrated in Figure 2.5.

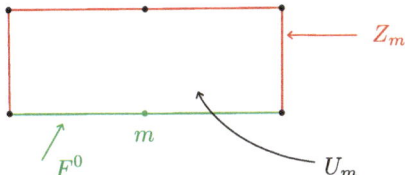

Figure 2.5: Open subset U_m and closed subset Z_m.

Definition 2.3.2. Let $A \subset P \cap M$ be an arbitrary subset of integral vectors in P. Let $V_A = \bigoplus_{m \in A} k e_m \subset H^0(X, L)$ be a linear system and $\varphi_A : X \to \mathbb{P}^{|A|-1}$ be the corresponding rational map.

  We denote by $\mathbb{R}_A$ the vector subspace of $\mathbb{R}^{1+r}$ generated by the vectors $(1, m)$, $m \in A$, and by $\mathbb{Z}_A = \mathbb{R}_A \cap \mathbb{Z}^{1+r}$ the corresponding saturated sublattice.

  We denote by $\mathbb{R}_P$ and $\mathbb{Z}_P$ the corresponding sets for $A = P \cap M$.

Theorem 2.3.3. *The following holds:*

1. *The rational map φ_A is regular, i.e., the base locus of the linear system V_A is empty if and only if $A \supset \mathrm{Vertices}(P)$. In this case, φ_A is a finite map of degree $|\mathbb{Z}_P : \langle (1, m), \ m \in A \rangle|$.*

2. *Assuming $A \supset \mathrm{Vertices}(P)$, the map φ_A is a closed embedding if and only if for every vertex v, the semigroup of integral vectors in the cone $\mathbb{R}_{\geq 0}(P - v) \subset M_{\mathbb{R}}$ is generated by the vectors $a - v$, $a \in A$.*

3. *In particular, if the semigroup $S_P = \mathrm{Cone}(1, P) \cap \mathbb{Z}^{1+r}$ is generated by the vectors $(1, m)$, $m \in A$, then φ_A is a closed embedding.*

Definition 2.3.4. We call a set $A \subset P \cap M$ *generating* if the group $\mathbb{Z}_P$ is generated by the vectors $(1, m)$, $m \in A$, and *totally generating* if the semigroup S_P is generated by $(1, m)$, $m \in A$. We call a lattice polytope *generating* (resp. *totally generating*) if the set $\mathrm{Vertices}(P)$ is generating (resp. totally generating).

  Thus, for a generating polytope, the map $\varphi_A : X \to \mathbb{P}^{|A|-1}$ is generically 1-to-1 for any $A \supset \mathrm{Vertices}(P)$ and, for a totally generating subset, φ_A is a closed embedding.

2.3.2 Linear systems on stable toric varieties

The following theorem is contained in [4]:

Theorem 2.3.5. *Let (X, L) be a stable toric variety. Then:*

1. $H^p(X, L) = 0$ *for $p > 0$;*

2. $H^0(X, L) = \bigoplus_{\rho^{-1}(M) \cap |\Delta|} k e_m$. *In other words, $H^0(X, L)$ is a direct sum of one-dimensional eigenspaces, one for each "integral" point of the topological space $|\Delta|$. Thus, $H^0(X, L)$ is the union of $H^0(X_i, L_i)$ for the irreducible components X_i, with subspaces corresponding to $X_i \cap X_j$ identified.*

2.4   Stable toric varieties over a projective variety V

2.4.1   Definition and main result

Let $\mathbb{P}^n$ be a projective space together with a T-linearized sheaf $\mathcal{O}(1)$. The linearization is the same as an assignment $z_j \mapsto m_j = \mathrm{wt}(z_j) \in M$, $j = 1, \ldots, n$.

Definition 2.4.1. A (stable) toric variety *over* $\mathbb{P}^{n-1}$ is a (stable) toric variety X together with a finite morphism $f \colon X \to \mathbb{P}^{n-1}$ and an isomorphism $L \simeq f^* \mathcal{O}(1)$ of T-linearized ample sheaves.

The homomorphism f is the same as a homomorphism of graded vector spaces $H^0(\mathbb{P}^{n-1}, \mathcal{O}(1)) = \bigoplus_{j=1}^n k z_j \to H^0(X, L)$. It gives a homomorphism

$$f^* \colon \bigoplus_{d \geq 0} H^0\big(\mathbb{P}^{n-1}, \mathcal{O}(d)\big) = k[z_1, \ldots, z_n] \longrightarrow R(X, L) = \bigoplus_{d \geq 0} H^0(X, L^d)$$

and the map $f \colon (X, L) \to (\mathbb{P}^{n-1}, \mathcal{O}(1))$ in the opposite direction. Thus, the morphism f is equivalent to picking n homogeneous eigenvectors $f^*(z_{m_j}) = e_{m_j} \in H^0(X, L)$ with $\mathrm{wt}(e_{m_j}) = m_j$.

Let $A \subset \{1, \ldots, m\}$ be the subset of those m_j such that $e_{m_j} \neq 0$. For each irreducible component X_i of X we get a set $A_i = A \cap P_i$. By the previous section, one has $A_i \supset \mathrm{Vertices}(P_i)$, otherwise the map f is not regular on X_i.

One can easily generalize the above definition by considering a T-invariant subvariety $V \subset \mathbb{P}^{n-1}$ with the sheaf $\mathcal{O}_V(1) = \mathcal{O}_{\mathbb{P}^{n-1}}|_V$.

Definition 2.4.2. A (stable) toric variety *over* $V \subset \mathbb{P}^{n-1}$ is a (stable) toric variety X together with a finite morphism $f \colon X \to V$ and an isomorphism $L \simeq f^* \mathcal{O}(1)$ of T-linearized ample sheaves.

The main result about stable toric varieties over $V \subset \mathbb{P}^{n-1}$ is the following:

Theorem 2.4.3 ([9]). *For each topological type $|\Delta|$ there exists a coarse moduli space $M_{|\Delta|}^T(V)$ of stable toric varieties over V. Further, $M_{|\Delta|}^T(V)$ is a projective scheme.*

Each point $[f \colon X \to V] \in M_{|\Delta|}^T(V)$ defines a tiling $\bigcup P_i$ of $|\Delta|$ by lattice polytopes, and the sets $A_i \supset \mathrm{Vertices}$. The points of $M_{|\Delta|}^T(V)$ with the same $\bigcup(P_i, A_i)$ form a locally closed stratum. This gives a stratification of $M_{|\Delta|}^T(V)$.

2.4.2   Moment map

When working over $\mathbb{C}$, there is an even nicer geometric connection between (X, L) and a lattice polytope P: there is a natural *moment map* $\mu \colon X(\mathbb{C}) \to M_{\mathbb{R}}$ whose image is P. So, $X(\mathbb{C})$ is fibered over the polytope P. Figure 2.6 gives an illustration for $(X, L) = (\mathbb{P}^1, \mathcal{O}(2))$.

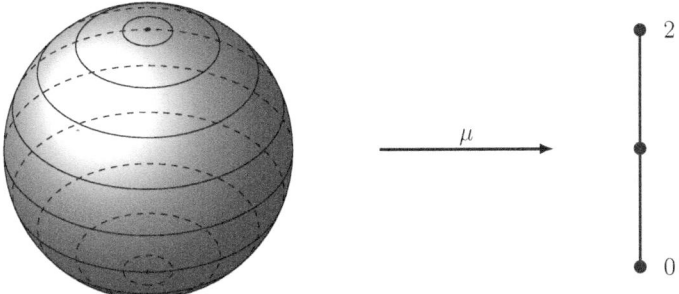

Figure 2.6: Moment map for $(\mathbb{P}^1, \mathcal{O}(2))$.

The moment map for $(\mathbb{P}^{n-1}, \mathcal{O}(1))$ with a T-linearized sheaf $\mathcal{O}(1)$ is defined by the formula

$$\mu(z_1, \ldots, z_n) = \frac{\sum |z_j|^2 \cdot m_j}{\sum |z_j|^2}.$$

The moment map for a (stable) toric variety over $\mathbb{P}^{n-1}$ is the composition $X(\mathbb{C}) \to \mathbb{P}^{n-1}(\mathbb{C}) \to M_\mathbb{R}$. Thus, if f is given by $f^* \colon z_j \mapsto c_j e_j \in H^0(X, L)$, where e_j is a homogeneous basis of $H^0(X, L)$, then the moment map $\mu \colon X(\mathbb{C}) \to M_\mathbb{R}$ is defined by the formula

$$\mu(p) = \frac{\sum |c_m e_m(p)|^2 \cdot m}{\sum |c_m e_m(p)|^2}, \qquad m \in |\Delta| \cap \rho^{-1}(M).$$

The preimage $\mu^{-1}(y)$ over a point in an a-dimensional face of P is isomorphic to the compact real torus $(S^1)^a$.

The moment map gives a nice representation of the T-orbits in X. The T-orbits are $\mu^{-1}(F^0)$, for all faces F in the tiling $\bigcup P_i$ of $|\Delta|$. Figure 2.7 gives an example of moment maps for the family of quadrics $x_0 x_2 = t x_1^2$ in $\mathbb{P}^2$ and its degeneration $x_0 x_2 = 0$.

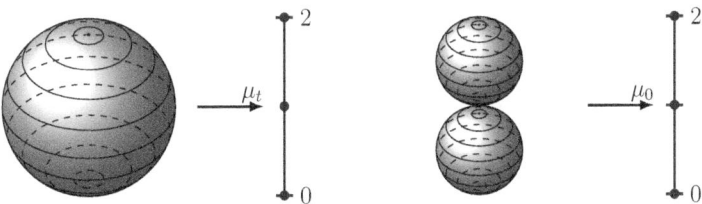

Figure 2.7: Moment map for $(\mathbb{P}^1, \mathcal{O}(2))$ and of its degeneration $\mathbb{P}^1 \cup \mathbb{P}^1$.

2.5 Stable toric pairs vs stable toric varieties over $\mathbb{P}^{n-1}$

Consider a toric variety (X, L) with a T-linearized sheaf L. Pick a homogeneous basis $H^0(X, L) = \bigoplus_{m \in P \cap M} k e_m$. Any section $s \in H^0(X, L)$ can be uniquely written as $s = \sum c_m e_m$. Let $A \subset P \cap M$ be the set of those m for which $c_m \neq 0$.

Theorem 2.5.1. *The following conditions are equivalent:*

1. $A \supset \mathrm{Vertices}(P)$;
2. *the rational map $X \mapsto \mathbb{P}^{n-1}$, $n = h^0(X, L)$, defined by $z_m \to c_m e_m$ is regular;*
3. *the divisor $D = (s)$ does not contain any T-orbits.*

Proof. We already saw the equivalence of 1. and 2.

To see the equivalence of 1. and 3., observe that condition (3) is equivalent to requiring that D does not contain any zero-dimensional T-orbits. The zero-dimensional orbits are in bijection with the vertices of P, $v \leftrightarrows Q_v$. For each vertex v of P, all the sections e_m with $m \neq v$ vanish at the point Q_v. So, $s(Q_v) \neq 0$ if and only if $c_v \neq 0$. $\qquad\square$

This shows that the moduli space $M_P^T(\mathbb{P}^{n-1})$ of stable toric varieties over $\mathbb{P}^{n-1}$ in this case is equivalent to the moduli space of stable toric *pairs* (X, D) of topological type P which satisfy condition (3) in the above theorem 2.5.1.

2.6 Singularities of stable toric varieties

2.6.1 Depth properties

It turns out that the singularities of a stable toric variety are completely determined by the topological space $|\Delta|$.

Definition 2.6.1. Let S be a topological space which has a structure of a finite simplicial complex. For a point $s \in S$, the link Link_s is the intersection of S with a small sphere centered at s.

The space S is called *Cohen–Macaulay* over the base field k if for all $s \in S$, one has $H_0(\mathrm{Link}_s, k) = k$ and $H_p(\mathrm{Link}_s, k) = 0$ for $0 < p < \dim \mathrm{Link}_s$.

Theorem 2.6.2. *The support $|\Delta|$ of a stable toric variety (X, L) is Cohen–Macaulay if and only if X is Cohen–Macaulay.*

Corollary 2.6.3. *If the support $|\Delta|$ of a stable toric variety (X, L) is a polytope, then X is Cohen–Macaulay.*

Proof. Indeed, in this case Link_s is either a sphere, or a disk of dimension $d = \dim X - 1$, so $H_0 = k$ and $H_p = 0$ for $0 < p < d$. $\qquad\square$

Example 2.6.4. The complex in Figure 2.8 is *not* Cohen–Macaulay, since the link at the origin is two closed intervals, and $H_0 = k^2$. This stable toric variety is a union of two normal surfaces glued at a single point. One can also see this directly: Cohen–Macaulay implies connected in codimension 1. The surface in Figure 2.8 is not connected in codimension 1.

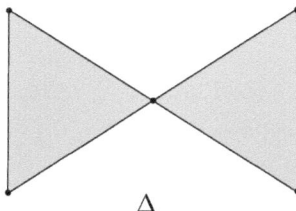

Δ

Figure 2.8: A non Cohen–Macaulay stable toric variety.

2.6.2 Log canonical and semi-log canonical

Toric varieties provide some of the easiest examples of log canonical singularities.

Lemma 2.6.5. *Let X be a toric variety and Δ be the union of the boundary divisors, the complement of the dense orbit. Then $K_X + \Delta \sim 0$ and the pair (X, Δ) is log canonical.*

Proof. The first property is well known, and the second property is a consequence of the first. Indeed, any toric variety has a toric resolution obtained by subdividing the cone. Let $f: Y \to X$ be such a resolution. Then Y is smooth and Δ^Y is a normal crossing divisor. Then,

$$f^*(K_X + \Delta) = f^*(0) = 0 = K_Y + \Delta^Y.$$

Since $\Delta^Y = \sum D_i$ is the sum of the boundary divisors with coefficients 1, (X, Δ) is lc. $\qquad\square$

What if we want a stable pair, i.e., it should be lc and $K + B$ should be ample? Then we need to add something ample to $K_X + \Delta$.

Lemma 2.6.6. *Let B be an ample effective divisor. Then*

1. $K_X + \Delta + \epsilon B$ *is ample for any $\epsilon > 0$;*
2. *the pair $(X, \Delta + \epsilon B)$ is lc for $0 < \epsilon \ll 1$ if and only if B does not contain any T-orbits, and if and only if B does not contain any zero-dimensional T-orbits.*

The proof of (2) follows by continuity: the only places where $(X, \Delta + \epsilon B)$ is not lc for $0 < \epsilon \ll 1$ are the places where we are already at the limit, i.e., the discrepancy is $a_D = -1$. These are precisely the boundary divisors and their intersections, i.e., the closures of the T-orbits. And the only closed T-orbits are zero-dimensional.

Similarly, stable toric varieties provide some of the easiest examples of semi log canonical singularities.

Lemma 2.6.7. *Let X be a stable toric variety whose topological type is a manifold with boundary (for example, a polytope). Let Δ be the union of its outside boundary divisors (as illustrated in Figure 2.9). Then,*

1. $K_X + \Delta = 0$ *and the pair (X, Δ) is slc;*
2. *for an effective divisor B, the pair $(X, \Delta + \epsilon B)$ is slc if and only if B does not contain any T-orbits, and if and only if B does not contain any zero-dimensional orbits.*

Indeed, normalizing reduces the situation to the toric case, and $\nu^*(K_X + \Delta_{\text{outside}}) = K_{X^\nu} + \Delta_{\text{all}}$.

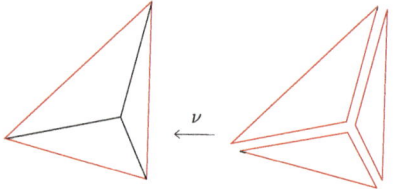

Figure 2.9: Outside boundary of a stable toric variety.

2.7 One-parameter degenerations

Many of the combinatorial constructions in this section originated in the work of Gelfand–Kapranov–Zelevinsky [23]. The conclusions from these combinatorial constructions differ in the following respect: [23] works with families of embedded cycles in $\mathbb{P}^{n-1}$, and some of these cycles may have multiplicities. However, the resulting family below is a family of stable toric varieties *over* $\mathbb{P}^{n-1}$ or, equivalently, the family of stable toric pairs (X_t, D_t) in which the varieties X_t are reduced, so there are no multiplicities.

Let us go in detail through a simple example, which hopefully illustrates everything there is to understand about one-parameter degenerations of stable toric varieties.

Consider a line $\mathbb{P}^1$ together with a divisor $\Delta + \epsilon B_t$, where $\Delta = P_0 + P_\infty$ is the boundary divisor, the complement of the dense torus orbit, and B_t is given by the equation

$$f_t = c_0 t^2\, x_0^5 + c_1 t\, x_0^4 x_1 + c_2\, x_0^3 x_1^2 + c_3\, x_0^2 x_1^3 + c_4 t\, x_0 x_1^4 + c_5 t\, x_1^5$$

for some fixed constants c_i. If all $c_i \neq 0$, then for any $t \neq 0$ the pair $(\mathbb{P}^1, \Delta + \epsilon B_t)$ is stable: it has lc singularities and ample $K_X + \Delta + \epsilon B_t$. Clearly, $\mathcal{O}_{\mathbb{P}^1}(B_t) \simeq \mathcal{O}_{\mathbb{P}^1}(5)$. It is a toric variety corresponding to the polytope $[0,5]$ in $M_{\mathbb{R}}$, $M = \mathbb{Z}$.

We would like to understand the limit of this pair as $t \rightsquigarrow 0$.

2.7.1 Complimentary degenerations

Let us associate to this family the following graph. For each of the points $m = 0, 1, \ldots, 5$ let $h(m)$ be the *height*, the valuation at t of the coefficient of $x_0^{5-m} x_1^m$ in f_t. This graph is shown in Figure 2.10.

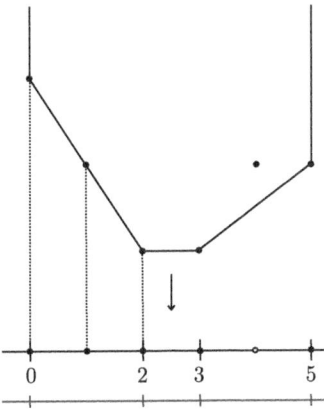

Figure 2.10: One-parameter degeneration of stable toric varieties over V.

The easiest way to degenerate the pair $(\mathbb{P}^1, B_t)$ is to simply look at the limit of the equation f_t as $t \to 0$. It is

$$f_t(0) = c_2 x_0^2 x_1^2 \big(c_2 x_0 + c_3 x_1 \big).$$

The pair $(\mathbb{P}^1, \Delta + \epsilon B_0)$ is not lc, since the coefficients of P_0 and P_∞ are $1 + 2\epsilon$. The log canonical model of $(\mathbb{P}^1, \Delta + \epsilon B_0)$ (see our Definition 1.2.5) is $(X_0^{(1)}, \Delta + \epsilon B_0^{(1)})$, where $X_0^{(1)} = \mathbb{P}^1$ and $B_0^{(1)} = (c_2 x_0 + c_3 x_1)$, which is a point distinct from $0, \infty$.

To be absolutely clear, $X_0^{(1)}$ does not correspond to the polytope $[0, 5]$ as the original toric variety $\mathbb{P}^1$. Instead, it corresponds to the polytope $[2, 3]$.

However, let us rescale the coordinates as follows: $y_0 = x_0$, $y_1 = t^{-1} x_1$. In the new coordinates, our family becomes

$$f_t = t^2 \big(c_0\, y_0^5 + c_1\, y_0^4 y_1 + c_2\, y_0^3 y_1^2 + c_3 t\, y_0^2 y_1^3 + c_4 t^3\, y_0 y_1^4 + c_5 t^4\, y_1^5 \big).$$

The height function is obtained from the previous one by adding a linear function $h_2(m) = h(m) - t - 2$. The limit now is

$$t^{-2} f_t(0) = y_0^3 (c_0 y_0^2 + c_1 y_0 y_1 + c_2 y_1^2).$$

Again, the pair $(\mathbb{P}^1, \Delta + \epsilon(t^{-2} f_t(0)))$ is not lc, since the coefficient of P_∞ is $1 + 2\epsilon$. Its log canonical model is $(X_0^{(2)}, \Delta + \epsilon B_0^{(2)})$, where $X_0^{(2)} = \mathbb{P}^1$ and $B_0^{(2)} = (c_0 y_0^2 + c_1 y_0 y_1 + c_2 y_1^2)$, which is two points distinct from $0, \infty$. The new variety $X_0^{(2)}$ corresponds to the polytope $[0, 2]$.

Finally, we can rescale as follows: $z_0 = x_0$, $z_1 = t x_1$. In the new coordinates, our family becomes

$$f_t = t^3 \big(c_0 t^5 z_0^5 + c_1 t^3 z_0^4 z_1 + c_2 t z_0^3 z_1^2 + c_3 z_0^2 z_1^3 + c_4 t z_0 z_1^4 + c_5 z_1^5 \big).$$

The height function is obtained from the previous one by adding a linear function $h_3(m) = h(m) + t - 3$. The limit now is

$$t^{-3} f_t(0) = z_1^3 (c_3 z_0^2 + c_5 z_1^2).$$

Again, the pair $(\mathbb{P}^1, \Delta + \epsilon(t^{-2} f_t(0)))$ is not lc, since the coefficient of P_0 is $1 + 3\epsilon$. Its log canonical model is $(X_0^{(3)}, \Delta + \epsilon B_0^{(3)})$, where $X_0^{(3)} = \mathbb{P}^1$ and $B_0^{(3)} = (c_0 y_0^2 + c_1 y_0 y_1 + c_2 y_1^2)$, which is two points distinct from $0, \infty$. The new variety $X_0^{(3)}$ corresponds to the polytope $[3, 5]$.

Clearly, the three degenerations above correspond to different choices of adjusting the height function by a linear homogeneous function $h(m) \mapsto h(m) + \ell(m)$ and then taking the minimum.

2.7.2   Stable degeneration

The polytope $P = [0, 5]$ lies in the space $M_\mathbb{R}$, where $M = \mathrm{Hom}(T, \mathbb{G}_m)$ is the character lattice of the torus acting on the varieties X_t. The polytope P gives a cone $\mathrm{Cone}(1, P)$ in $(\mathbb{Z} \oplus M) \otimes \mathbb{R}$, and each variety X_t, $t \neq 0$, can be written as the Proj of the semigroup algebra $k[\mathrm{Cone}(1, P) \cap (\mathbb{Z} \oplus M)]$.

Now, to the family X_t let us associate an infinite polyhedron lying in the space $M \oplus \mathbb{Z}$. The additional $\mathbb{Z}$ corresponds to the height and not to the degree.

Definition 2.7.1. The polyhedron P^+ is the lower convex envelope of the rays $(m, h(m) + \mathbb{R}_{\geq 0})$. It is depicted in Figure 2.10, and is semi-infinite in the upward direction.

Consider a cone over $(1, P^+)$ lying in the space $(\mathbb{Z} \oplus M \oplus \mathbb{Z}) \otimes \mathbb{R}$. Let R^+ be the semigroup algebra

$$R^+ = k\big[\mathrm{Cone}(1, P^+) \cap (\mathbb{Z} \oplus M \oplus \mathbb{Z}) \big].$$

A point (d, m, h) corresponds to a monomial $t^h x^{(d,m)}$ of degree $d \geq 0$. Let $X^+ = \operatorname{Proj} R^+$. Since R^+ is a $k[t]$-algebra, X^+ has a natural morphism $f \colon X^+ \to \operatorname{Spec} k[t] = \mathbb{A}^1_t$. This is our degenerating family.

The central fiber of this family is a scheme with three irreducible components $X^{(1)}$, $X^{(2)}$, $X^{(3)}$, corresponding to the lower faces of P^+. Projecting these faces down to $[0, 5]$ gives a polyhedral subdivision of it into the intervals $[0, 2]$, $[2, 3]$ and $[3, 5]$.

Definition 2.7.2. For any lattice polytope P, a polyhedral subdivision obtained by projecting down the lower envelope of the polyhedron P^+ for some height function $h \colon P \cap M \to \mathbb{R}$ is called a *convex subdivision*. Other names used for such subdivisions are *regular*, or *coherent*.

Remark 2.7.3. Not every polyhedral subdivision is convex. The standard counterexample is depicted in Figure 2.11.

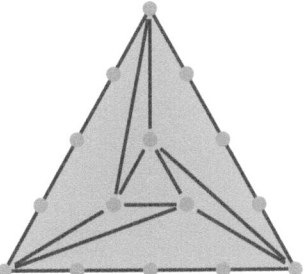

Figure 2.11: A non-convex tiling.

The corresponding stable toric variety does not appear as the limit of a one-parameter family of $(\mathbb{P}^2, D_t)$ with $\mathcal{O}_{\mathbb{P}^2}(D_t) \simeq \mathcal{O}_{\mathbb{P}^2}(4)$.

To continue with our construction, there is yet another twist: the central fiber of $\operatorname{Proj} R^+ \to \mathbb{A}^1_t$ is not reduced, because the component X_3 appears it it with multiplicity 2. To see this, note that the monomial $t x_0 x_1^4 \in R^+$ does not lie in the ideal (t), but its square does: $t^2 x_0^2 x_1^8 = t \cdot x_0^2 x_1^3 \cdot t x_1^5$. The reason for this is that the set $\{3, 5\}$ of the polytope $[3, 5]$ is not generating, the vertices $(1, 3)$, $(1, 5)$ generate a sublattice of index 2 in $\mathbb{Z} \oplus M$.

This changes, however, by making a finite ramified base change $\mathbb{A}^1_s \to \mathbb{A}^1_t$, $t = s^2$. After this base change the central fiber is a reduced seminormal union of the toric varieties $X^{(1)}$, $X^{(2)}$, $X^{(3)}$, each isomorphic to $\mathbb{P}^1$. No further finite base change changes the central fiber.

The original equation f_t defines a section of the invertible sheaf $\mathcal{O}(1)$ on $X = \operatorname{Proj} R^+$, a relative Cartier divisor B on X which restricts to the divisor $B_0 = B_0^{(1)} \cup B_0^{(2)} \cup B_0^{(3)}$ on the central fiber X_0. Moreover, some thinking about this will convince the reader that R^+ is *the only* graded subalgebra of $k[M \oplus \mathbb{Z}]$ for which f_t stays a regular section of $\mathcal{O}(1)$ and satisfies the following condition on the central fiber: "B_0 does not contain any T-orbits". This condition was part of our definition of stable toric varieties; so, the moduli functor of stable toric pairs is proper.

2.7.3 Maximal and higher codimension degenerations

What will happen if we rescale the height function by $h(m) \mapsto h(m) + \frac{1}{2}m$? For this to make sense, we will have to first make the base change $t = s^2$. The new height function will be $h'(m) = 2h(m)$, and then we will rescale it by $h'(m) \mapsto h'(m) + m$.

The minimum of $h'(m)$ is achieved at the unique point $\{2\}$. The limit of the divisors B_t is given by the equation $x_0^3 x_1^2$. The corresponding pair is $(\mathbb{P}^1, (1 + 3\epsilon)P_0 + (1 + 2\epsilon)P_\infty)$ and it is not lc.

If we attempt to find its log canonical model, then it is not going to work because the "round-down" pair $(\mathbb{P}^1, P_0 + P_\infty)$ is not of general type, the divisor $K_{\mathbb{P}^1} + P_0 + P_\infty$ is not of general type. The image of its Iitaka fibration, however, is a point.

Other "non-maximal" choices give the toric varieties, each a single points, corresponding to the polytopes $\{0\}$, $\{3\}$, and $\{5\}$. Clearly, the variety X is glued from the three irreducible components $X_0^{(1)}$, $X_0^{(2)}$, $X_0^{(3)}$ along these smaller-dimensional varieties.

How do we recognize if a certain choice of a height function is going to give us an irreducible component in the central fiber or a smaller-dimensional stratum of the limit variety X_0?

In terms of the graph, the answer is obvious: it is whether the corresponding polytope is maximal-dimensional or not.

In terms of the pair, the answer is as follows: the maximal-dimensional degenerations correspond to the pairs $(X, \Delta + \epsilon B)$ for which the automorphism group is finite. For $X_0^{(1)}$, $X_0^{(2)}$, $X_0^{(3)}$ the groups are 1, 1, and $\mu_2 \simeq \mathbb{Z}_2$. For the degenerations giving lower-dimensional strata of the central fiber X_0, the group contains $\mathbb{G}_m$ and is infinite. The dimension of the group equals the codimension of the stratum.

2.8 Toric varieties associated to hyperplane arrangements

A hyperplane arrangement is a collection of n hyperplanes $B_1, \dots, B_n$ in a projective space $\mathbb{P}^{r-1}$; these hyperplanes are allowed to coincide. We consider them up to an isomorphism of pairs, i.e., $(\mathbb{P}^{r-1}, B_1, \dots, B_n)$ is isomorphic to $(\mathbb{P}^{r-1}, B_1', \dots, B/n)$ if and only if there exists an automorphism $g \in \mathrm{PGL}(r)$ of $\mathbb{P}^{r-1}$ such that $g(B_1) = B_1', \dots, g(B_n) = B_n'$.

We will be concerned with complete moduli of hyperplane arrangements. On the boundary of this moduli space the projective space $\mathbb{P}^{r-1}$ will split up somehow and degenerate to some non-normal variety, a higher-dimensional analogue of a stable curve.

The general idea is very simple. Since we understand so well degenerations of toric varieties, let us associate to a hyperplane arrangement $(\mathbb{P}^{r-1}, B_1, \ldots, B_n)$ a toric pair $(Y, \Delta + \epsilon D)$ or a toric variety $Y \to V$ over some projective variety. Let us do this in a reversible way, so that we can go back from Y to $(\mathbb{P}^{r-1}, B_1, \ldots, B_n)$.

Then, degenerations of toric varieties will give us degenerations of hyperplane arrangements, and the complete moduli spaces of stable toric varieties will give us complete moduli spaces of stable hyperplane arrangements.

2.8.1 The Gelfand–MacPherson correspondence

There are two dual ways to work with hyperplane arrangements, related by the Gelfand–MacPherson correspondence, which we explain now.

Consider an $(r \times n)$-matrix A of rank r with nonzero columns.

$$A = \begin{pmatrix} a_{11} & a_{12} & \cdots & a_{1n} \\ a_{21} & a_{22} & \cdots & a_{2n} \\ \cdots & \cdots & \cdots & \cdots \\ a_{r1} & a_{r2} & \cdots & a_{rn} \end{pmatrix}.$$

The columns of this matrix, considered as linear functions $f_i(x) = a_{1i}x_1 + \cdots + a_{r1}x_r$, define n hyperplanes $B_1, \ldots, B_n$ in a projective space $\mathbb{P}^{r-1}$. The condition $\operatorname{rank} A = r$ is equivalent to the following condition, which we want to stress:

$$\bigcap_{i=1}^{n} B_i = \varnothing.$$

Let $\operatorname{Mat}^0(r, n)$ be the set of all such matrices. Let $\mathbb{P}(\operatorname{Mat}^0(r, n))$ be the corresponding projective space of dimension $rn - 1$. The set $\operatorname{HA}(r, n)$ of isomorphism classes of n hyperplanes in $\mathbb{P}^{r-1}$ is the quotient

$$\operatorname{HA}(r, n) = \operatorname{PGL}(r) \setminus \mathbb{P}(\operatorname{Mat}^0(r, n)) / T,$$

where $\operatorname{PGL}(r) = \operatorname{GL}(r)/\mathbb{G}_m$ and $T = (\mathbb{G}_m^n)/\operatorname{diag}\mathbb{G}_m$ is a torus of dimension $n - 1$. The group $\operatorname{PGL}(r)$ acts on the rows by changing the basis in $\mathbb{A}^r$, and the torus T acts by scalar multiplication on the columns, rescaling the equations without changing the hyperplanes B_i they define.

Now, if we take the quotient $\mathbb{P}(\operatorname{Mat}^0(r, n))/T$ first, we obtain $(\mathbb{P}^{r-1\vee})^n$, the set of hyperplanes in a fixed projective space $\mathbb{P}^{r-1}$. Then

$$\operatorname{HA}(r, n) = \operatorname{PGL}(r)\setminus(\mathbb{P}^{r-1\vee})^n.$$

If we take the quotient $\mathrm{PGL}(r)\backslash\mathbb{P}(\mathrm{Mat}^0(r,n))$ first, we obtain $\mathrm{G}^0(r,n)$, the open subset of the Grassmannian $\mathrm{G}(r,n)$ of r-dimensional quotient spaces $\mathbb{A}^n \to V^*$ of a fixed n-dimensional space with the additional condition that V are not contained in any of the n coordinate hyperplanes. (Alternatively, $\mathrm{G}(r,n)$ parameterizes the subspaces $V \subset \mathbb{A}^n$, but we treat the columns as linear equations, so the interpretation with the quotients suits us better.) Then $\mathrm{HA}(r,n) = \mathrm{G}^0(r,n)/T$.

So a single hyperplane arrangement, up to an isomorphism, is the same as a T-orbit inside $\mathrm{G}^0(r,n)$. A point in the Grassmannian is $\mathbb{A}^n \to V^*$, i.e., $\mathbb{P}V \subset \mathbb{P}^{n-1}$. The hyperplanes B_i are the intersections of the n coordinate hyperplanes $H_i = \{z_i = 0\} \subset \mathbb{P}^{n-1}$ with $\mathbb{P}V$. The torus $T = \mathbb{G}_m^n / \operatorname{diag} \mathbb{G}_m$ acts by rescaling the n homogeneous coordinates z_i.

Let $T \cdot [V]$ be a single orbit. Its closure $Y = \overline{T \cdot [V]}$ is then a projective toric variety. A priori, it may be non-normal. It turns out, however, that the Grassmannians are very special and nice, and Y is indeed an ordinary normal toric variety (see Theorem 4.1.6).

Note that the orbit does not have a special "origin", so Y is a toric variety and not a torus embedding. It is a toric variety $Y \to \mathrm{G}(r,n)$ over the Grassmannian.

In order to recover the hyperplane arrangement from Y, let $P \to \mathrm{G}(r,n)$ be the universal family, $P \subset \mathbb{P}^{n-1} \times \mathrm{G}(r,n)$, whose fiber over $[\mathbb{P}V \subset \mathbb{P}^{n-1}]$ is $\mathbb{P}V \subset \mathbb{P}^{n-1}$. If Y^0 is the dense T-orbit of Y and

$$P_{Y^0} := p_2^{-1}(Y^0) = P \times_{\mathrm{G}(r,n)} Y^0 \subset P,$$

then the hyperplane arrangement is P_{Y^0}/T.

So, the set $\mathrm{HA}(r,n)$ is the quotient set $\mathrm{G}(r,n)/T$, and the hyperplane arrangements are the T-quotients of the preimages of these orbits in the universal family $P \to \mathrm{G}(r,n)$.

Of course, the quotient set $\mathrm{G}(r,n)/T$ is extremely nasty and has no algebraic variety structure. To get a nice space, we will have to take the GIT quotient. To recover the hyperplane arrangements themselves, we will have to take the GIT quotients of projective varieties $P_Y = P \times_{\mathrm{G}(r,n)} Y$. So we will need to understand the issues involved with taking such quotients.

The degenerations of hyperplane arrangements will correspond to degenerations of toric varieties, which will be some stable toric varieties over $\mathrm{G}(r,n)$. Thus, irreducible components X_i of a degeneration $X = \bigcup X_i$ will correspond to some toric varieties $Y_i \subset \mathrm{G}(r,n)$, which in turn means that they will correspond to some new hyperplane arrangements $[\mathbb{P}V_i \subset \mathbb{P}^{n-1}]$. So, somehow, a limit of a family of hyperplane arrangements will be glued from several other hyperplane arrangements.

2.8.2 Torus action on the Grassmannian

Let us explain this torus action in more detail. There is a natural action of the torus $\widetilde{T} = (\mathbb{G}_m)^n$ on $\mathbb{A}^n$:

$$(\lambda_1, \ldots, \lambda_n) \cdot (z_1, \ldots, z_n) = (\lambda_1 z_1, \ldots, \lambda_n z_n).$$

This defines an action of $\widetilde{T}$ on the Grassmannian $G(r, n)$, as follows. If $G(r, n) \subset \mathbb{P}^N$, $N = \binom{n}{r} - 1$, is the Plücker embedding with Plücker coordinates p_I for all $I \subset \overline{n}$, $|I| = r$, then the induced action is

$$(\lambda_1, \ldots, \lambda_n) \cdot p_I = \left(\prod_{i \in I} \lambda_i \right) p_I.$$

An algebraic action of a torus on any vector space A is diagonalizable, and one gets a decomposition $A = \bigoplus_{\chi \in \Lambda_{\widetilde{T}}} A_\chi$ into eigenspaces. Here, $\Lambda_{\widetilde{T}} = \mathrm{Hom}(\widetilde{T}, \mathbb{G}_m) = \mathbb{Z}^n$ is the character group of $\widetilde{T}$. Thus, to every eigenvector v one assigns a character, also called its *weight* $\mathrm{wt}(v) \in \mathbb{Z}^n$.

In these terms, one has $\mathrm{wt}(z_i) = e_i$ and $\mathrm{wt}(p_I) = e_I = \sum_{i \in I} e_i$. Since $\mathrm{diag}\,\mathbb{G}_m \subset \widetilde{T}$ acts trivially on $\mathbb{P}^{n-1}$, the torus $T = \widetilde{T}/\mathrm{diag}\,\mathbb{G}_m \simeq \mathbb{G}_m^{n-1}$ also acts on $\mathbb{P}^{n-1}$ and $G(r, n)$. The character group of T is

$$\Lambda_T = \left\{ \sum n_i e_i \mid \sum n_i = 0 \right\}.$$

We do not have a natural T-action on $\mathbb{A}^n$, so there are no weights in Λ_T assigned to the homogeneous coordinates z_i, p_I. However, for the coordinates on the standard affine covers one has $\mathrm{wt}(z_i/z_j) = e_i - e_j$ and $\mathrm{wt}(p_I/p_J) = e_I - e_J$.

2.8.3 Moment polytope of a hyperplane arrangement

Under the Plücker embedding $Y \subset G(r, n) \subset \mathbb{P}^{N-1}$, the moment polytope of the toric variety $Y = \overline{T \cdot [V]}$ is the convex hull of the vectors e_I for all $I \subset \overline{n}$, $|I| = r$, such that the corresponding Plücker coordinate $p_I(V)$ of the space $V \subset \mathbb{A}^n$ is nonzero.

This condition is equivalent to any of the following two conditions:

1. $\bigcap_{i \in I} B_i = \varnothing$ in $\mathbb{P}V$;
2. the linear equations f_i of the hyperplanes B_i form a basis in the dual space V^*.

This moment polytope is called a *matroid polytope*. The collection of vectors $f_i \in V^*$ is called a *vector matroid*.

Thus, to understand the compactified moduli spaces of hyperplane arrangements, we will have to understand matroid polytopes, and matroids in general.

Chapter 3

Matroids

A matroid is a pair $M = (E, \mathcal{I})$ consisting of a (usually finite) set E and a set $\mathcal{I} \subset 2^E$ of subsets called the *independent sets*. Equivalently, it can be defined using *bases*, or using the *rank function* $r: 2^E \to \mathbb{Z}_{\geq 0}$.

Usually, we identify E with the set $\bar{n} = \{1, \ldots, n\}$. For us, the only interesting matroids are *vector matroids*, also called *representable* or *linear* matroids. They are very easy to understand.

There are many introductory books on matroids. Some standard sources include [50, 53, 63]. All of the facts that we state without proof can be found there.

3.1 Vector (or representable) matroids

3.1.1 Vector matroids using independent sets

Fix a field k. Consider n vectors $f_1, \ldots, f_n$ spanning a k-vector space W of dimension r. Call a subset $I \subset \bar{n}$ an *independent set* if the vectors $\{f_i, \ i \in I\}$ are linearly independent.

Definition 3.1.1. The *vector matroid* represented by vectors $f_1, \ldots, f_n \in W$ is the pair $M = (\bar{n}, \mathcal{I})$, where $\mathcal{I}$ is the set of all independent sets. The *rank* r of M is the dimension of the span $\langle f_i \rangle$.

Any undergraduate student who had a first course in linear algebra should have no trouble proving the following lemma.

Lemma 3.1.2. *The set $\mathcal{I}$ satisfies the following properties:*

1. $\emptyset \in \mathcal{I}$ *(for some, this could be a matter of convention);*
2. *(monotonicity) if I is independent and $J \subset I$, then J is independent;*
3. *(independent set exchange property) if I, J are independent and $|I| > |J|$, then there exists $i \in I \setminus J$ such that $J \cup \{i\}$ is independent.*

Vectors are allowed to be zero. Of course, the zero vectors never appear in any independent set, so they are not very interesting and can be ignored for most purposes.

Definition 3.1.3. A vector matroid is *loopless* if all vectors f_i are nonzero.

The name *loop* for a zero vector comes from graphs, another major source of matroids, see Section 3.10.4.

3.1.2   Vector matroids using bases

Another, more economical way to define matroids is using bases. Call a subset $I \subset \overline{n} = \{1, \ldots, n\}$ a *base* if the vectors $\{f_i, \ i \in I\}$ form a basis of W.

Definition 3.1.4. The vector matroid represented by the vectors $f_1, \ldots, f_n \in W$ is the pair $(\overline{n}, \mathcal{B})$, where $\mathcal{B}$ is the set of all bases.

Again, the following is an elementary fact.

Lemma 3.1.5. *Let $\mathcal{B}$ be a matroid on the set $\overline{n}$. Then*

1. *(Base exchange property) For two bases I, J and $i \in I \setminus J$, there exists $j \in J \setminus I$ such that $I \setminus \{i\} \cup \{j\} \in \mathcal{B}$.*

Of course, it is easy to go from $\mathcal{I}$ to $\mathcal{B}$ and back: the bases are the maximal independent sets, and independent sets are arbitrary subsets of bases, including $\varnothing$.

Example 3.1.6. Let $\mathcal{B}(r, n)$ be the set of *all* subsets of $\overline{n}$ with cardinality r. If the field k is large enough with respect to n (for example infinite), then $\mathcal{B}(r, n)$ is a vector matroid over k, called *uniform matroid*. To construct it, just take n vectors in k^r in general position, with no linear dependencies between any subset of up to r of them.

3.1.3   Vector matroids using the rank function

Define the following function on the set of subsets of $\overline{n}$

$$
\begin{aligned}
r \colon 2^{\overline{n}} &\longrightarrow \mathbb{Z}_{\geq 0}, \\
I &\longmapsto \dim\langle f_i, \ i \in I\rangle.
\end{aligned}
$$

Here, $\langle f_i, \ i \in I\rangle$ denotes the span of the vectors f_i. The following facts are once again elementary.

Lemma 3.1.7. *The following holds:*

1. *for any set $I \subset \overline{n}$, one has $r(I) \leq |I|$;*
2. *(monotonicity) if $I \subset J$ then $r(I) \leq r(J)$;*
3. *(submodularity) $r(I \cup J) + r(I \cap J) \leq r(I) + r(J)$.*

Of course, it is easy to go from independent sets to the rank function and back. For any subset $J \subset \bar{n}$, its rank is the cardinality of the largest independent subset $I \subset J$. Vice versa, given the rank function we can recognize the independent sets as those satisfying $r(I) = |I|$.

3.1.4 Other characterizations of vector matroids

There are other equivalent ways to characterize vector matroids, which we mention without going into details: using spanning sets, using circuits (minimal dependent sets), using the span (or closure) operation on $2^{\bar{n}}$, etc.

3.1.5 Vector matroids and hyperplane arrangements

For geometric reasons, we should be switching to the dual picture of hyperplane arrangements. From now on, all our matroids will be loopless.

Let $V = W^*$ be the dual space, and think of the vectors $f_i \in W = V^*$ as nonzero linear functions on V. Each of them defines a hyperplane $B_i \subset \mathbb{P}V \simeq \mathbb{P}^{r-1}$. Note that the condition $f_i \neq 0$ assures that B_i is actually a divisor. Also, the condition that f_i generate V^* is equivalent to $B_1 \cap \cdots \cap B_n = \varnothing$. We will assume both conditions from now on.

For convenience, let us introduce the notation for the following projective linear subspace of $\mathbb{P}V$: $B_I = B(I) := \bigcap_{i \in I} B_i$. Here is the translation of the notions we introduced above into the language of hyperplane arrangements:

Lemma 3.1.8. *The following holds:*

1. $r(I) = \operatorname{codim} B(I)$, *so* $B(I) \simeq \mathbb{P}^{r-1-r(I)}$;
2. I *is independent if and only if* $\operatorname{codim} B(I) = |I|$;
3. I *is a base if and only if* $B(I) = \varnothing$ *and* $|I| = r = \dim \mathbb{P}V + 1$.

(By convention, we set $\operatorname{codim} \varnothing = r$ *and* $\mathbb{P}^{-1} = \varnothing$.)

3.2 Abstract matroids

3.2.1 Definitions

The notion of an abstract matroid merely captures the abstract properties of vector matroids listed in Lemmas 3.1.2, 3.1.5, and 3.1.7. Below, we give three equivalent definitions. One can easily go from one definition to the other in the same way as we did for vector matroids.

Definition 3.2.1. A *matroid* M is a pair $(E, \mathcal{I})$ of a set E and a nonempty set $\mathcal{I} \subset 2^E$ of subsets of E, called *independent sets*, satisfying the following properties:

1. $\varnothing \in \mathcal{I}$;
2. (monotonicity) if I is independent and $J \subset I$, then J is independent;

3. (independent set exchange property) if I, J are independent and $|I| > |J|$, then there exists $i \in I \smallsetminus J$ such that $J \cup \{i\}$ is independent.

Definition 3.2.2. A *matroid* M is a pair $(E, \mathcal{B})$ of a set E and a set $\mathcal{B} \subset 2^E$ of subsets of E, called *bases*, satisfying the following property:

1. (Base exchange property) For two bases I, J and $i \in I \smallsetminus J$, there exists $j \in J \smallsetminus I$ such that $I \smallsetminus \{i\} \cup \{j\} \in \mathcal{B}$.

Definition 3.2.3. A *matroid* M is a pair (E, r) of a set E and a nonnegative function $r : 2^E \to \mathbb{Z}_{\geq 0}$ on the set of subsets of $\overline{n}$ satisfying the following properties:

1. for any set $I \subset E$, one has $r(I) \leq |I|$;
2. (monotonicity) if $I \subset J$, then $r(I) \leq r(J)$;
3. (submodularity) $r(I \cup J) + r(I \cap J) \leq r(I) + r(J)$.

3.2.2   Non-representable matroids

All matroids of ranks 1 and 2 are representable over any infinite field. In higher rank, however, this is not true in general. Here is the smallest and simplest example. Consider the 7 points and 7 lines of the finite projective plane $\mathbb{P}^2(\mathbb{F}_2)$ over the field $\mathbb{F}_2$ with two elements (the Fano plane). This configuration defines the *Fano matroid* of rank 3 on $\overline{7}$, which is nonrealizable over any field of characteristic different from 2.

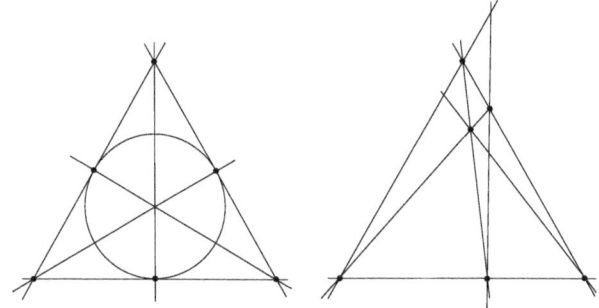

Figure 3.1: Fano and non-Fano matroids.

On the other hand, the *non-Fano matroid* is obtained by removing the line that looks like a circle in the picture and replacing it with an ordinary line passing only through two of the points. Non-Fano is realizable over any field of characteristic different from 2, but it is *not* realizable in characteristic 2 because in the latter case the 7th line *must* pass through the third point.

To construct a matroid which is not realizable over *any* field at all, it is sufficient to take the union of the two examples. This is a matroid of rank 3 on the set $\overline{14}$ whose restriction to $S_1 = \{1, \ldots, 7\}$ is a Fano matroid and to $S_2 = \{8, \ldots, 14\}$ is a non-Fano matroid. Otherwise, we declare the lines to be in general position. Thus, the bases are of the form $I_1 \sqcup I_2$, $|I_1| + |I_2| = 3$, where the sets $I_i \subset S_i$ are independent.

The matroid with the smallest n (but not the smallest r) which is not representable over any field is the Vámos matroid. It has $n = 8$ and $r = 4$. The eight elements can be pictured as the eight vertices of a cube. The minimal dependent sets are the six faces of the cube except for the parallel faces 1234 and 5678, plus the set 1537 joining the opposite parallel edges. All together, there are five minimal dependent sets.

One can read more about non-representable matroids in [51].

3.3 Connected components of a matroid

3.3.1 Definition

Consider a vector matroid represented by some vectors $f_1, \ldots, f_n$ in $W = V^*$ on the set $E = \overline{n}$. Now suppose that $W = W_1 \oplus W_2$ with $\dim W_i \geq 1$ and that the vectors are split into two groups, $E_1 \sqcup E_2$, so that for $i \in E_s$ one has $f_i \in W_s$. It follows that

1. a set $I \subset E$ is independent if and only if $I = I_1 \sqcup I_2$, and $I_s \subset E_s$ are independent sets;

2. $r(E_1) + r(E_2) = r(E)$;

3. more generally, for any $I \subset E$ one has $r(I \cap E_1) + r(I \cap E_2) = r(I)$.

In this situation one says that the vector matroid is *decomposable* or *disconnected*. In the opposite case, when there is no such decomposition, the matroid is called *indecomposable* or *connected*.

For an abstract matroid on the set E, the definition is the same: the matroid is decomposable if and only if there is a decomposition $E = E_1 \sqcup E_2$ satisfying any of the three properties above. (It is an exercise to check that they are equivalent.)

By repeating the process, we eventually get a decomposition $E = E_1 \sqcup \cdots \sqcup E_p$ for which the matroids M_s on the sets E_s are connected. It can be proven that, for any matroid, such a decomposition is unique. The matroids M_s are called *connected components* of M.

3.3.2 Geometric meaning for hyperplane arrangements

Let $f_1, \ldots, f_n \in V^*$, $f_i \neq 0$, be a loopless vector matroid of rank $r = \dim V$. The corresponding hyperplane arrangement consists of n hyperplanes B_i in $\mathbb{P}V \simeq \mathbb{P}^{r-1}$.

Let $V = V_1 \oplus V_2$, $n = n_1 + n_2$, $r = r_1 + r_2$, and suppose that the matroid is decomposable. Then we get two hyperplane arrangements, namely n_1 hyperplanes in $\mathbb{P}V_1 \simeq \mathbb{P}^{r_1-1}$ and n_2 hyperplanes in $\mathbb{P}V_2 \simeq \mathbb{P}^{r_2-1}$, and our original hyperplane arrangement is their *join*.

Below are the pictures of decomposable hyperplane arrangements for $r = 2$ and $r = 3$. The corresponding partitions of the rank are $2 = 1 + 1$, $3 = 1 + 1 + 1$, and $3 = 2 + 1$. The "elementary blocks" are hyperplane arrangements on $\mathbb{P}^0$ and $\mathbb{P}^1$.

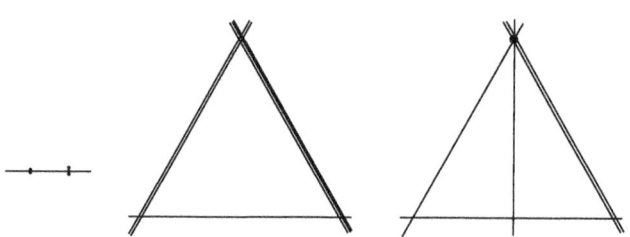

Figure 3.2: Decomposable matroids.

An automorphism of a hyperplane arrangement is an automorphism $g \colon \mathbb{P}V \to \mathbb{P}V$ such that $g(B_i) = B_i$ for each $i \in \overline{n}$.

The automorphism group of a connected hyperplane arrangement is trivial. Indeed, pick r of the vectors f_i that form a basis of V^*. Without loss of generality, we can assume that these are $f_1, \ldots, f_r$ and they are the standard basis vectors of k^n. Then $\mathrm{Aut}(\mathbb{P}V, B_1, \ldots, B_r)$ is the torus $T = (k^*)^r / \operatorname{diag} k^*$. For each additional vector f_i, let $S_i \subset \overline{r}$ be the set of indices for which the sth coordinate of f_i is nonzero. Then the subgroup of T sending B_i to itself consists of the elements $(\lambda_1, \ldots, \lambda_r)$ such that λ_s are all the same for $i \in S_i$. It is easy to see that the matroid is connected if and only if $\bigcup_{i=r+1}^{n} S_i = \overline{r}$. In this case, we obtain $\mathrm{Aut}(\mathbb{P}V, B_1, \ldots, B_n) = 1$.

For a general hyperplane arrangement, one has a decomposition $V = V_1 \oplus \cdots \oplus V_c$. The automorphism group consists of the dilations in each of the linear spaces V_i. The global dilations of V act trivially on $\mathbb{P}V$, so $\mathrm{Aut}(\mathbb{P}V, B_1, \ldots, B_n) = (k^*)^c / \operatorname{diag} k^* \simeq (k^*)^{c-1}$. Thus, the number of connected components of a hyperplane arrangement is easy to recognize geometrically.

Remark 3.3.1. It is *not true* that any hyperplane arrangement with a trivial automorphism group contains r hyperplanes in general position. A counterexample is provided by the columns of the following matrix, defining $(\mathbb{P}^3, B_1, \ldots, B_6)$:

$$\begin{pmatrix} 1 & 1 & 1 & 0 & 0 & 0 \\ 1 & 0 & 0 & 1 & 1 & 0 \\ 0 & 1 & 0 & 1 & 0 & 1 \\ 0 & 0 & 1 & 0 & 1 & 1 \end{pmatrix}.$$

However, the statement is true for hyperplane arrangements in $\mathbb{P}^1$ and $\mathbb{P}^2$.

3.4 Matroids of rank 1

A matroid of rank 1 is simply a set of n elements about which only one thing is important: which ones are linearly dependent as 1-element sets (loops) and which ones are linearly independent (non-loops); one must have at least one non-loop. Obviously, it is representable over any field. Thus, all loopless rank-1 matroids on n elements are isomorphic.

The corresponding hyperplane arrangement is $\mathbb{P}^0$ together with n divisors $B_1, \ldots, B_n$, so $B_i = \varnothing$. Despite being so trivial, this matroid becomes meaningful when we take joins with other hyperplane arrangements, as in Section 3.3. In the higher-dimensional projective space it becomes a nonempty hyperplane of multiplicity n.

3.5 Matroids of rank 2

All matroids of rank 2 are vector matroids, and can be represented over any infinite field.

Let $\mathcal{B}$ be a loopless vector matroid of rank 2. The corresponding hyperplane arrangement is a collection of p points B_i on $\mathbb{P}V = \mathbb{P}^1$. The condition $\cap B_i = \varnothing$ means that $p \geq 2$. Some points may coincide. Let $Q_1, \ldots, Q_p$ be the distinct points, and consider the partition $\overline{n} = J_1 \sqcup \cdots \sqcup J_p$ given by

$$i \in J_s \iff P_i = Q_s.$$

A pair (i, j) is *not* a base if i and j lie in the same J_s; all other pairs are bases.

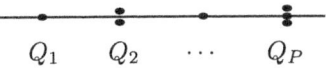

$$Q_1 \qquad Q_2 \qquad \cdots \qquad Q_P$$

Figure 3.3: Matroids of rank 2.

The matroid is connected if and only if it has at least three points. The decomposable matroids correspond to a partition $n = n_1 + n_2$ with $n_i \geq 1$; there are $\lfloor n/2 \rfloor$ of them.

The indecomposable matroids correspond to partitions of n into three or more parts. Up to the action of the permutation group S_n, there are $p_{\geq 3}(n) = p(n) - \lfloor n/2 \rfloor - 1$ of them, where $p(n)$ is the partition function. The values for low n are shown in Table 3.1.

We see that matroids of rank 2 are extremely simple. This is one reason why moduli spaces of stable curves of genus 0 are so nice and smooth and relatively simple. Unfortunately, that is where the simplicity ends. Matroids of rank 3 are already very complicated for larger n. And so are the moduli spaces of stable surfaces.

n	formula	3	4	5	6	7	8	9
decomposable	$\lfloor n/2 \rfloor$	1	2	2	3	3	4	4
connected	$p_{\geq 3}(n)$	1	2	4	7	11	17	25
total	$p(n) - 1$	2	4	6	10	14	21	29

Table 3.1: Number of connected rank-2 matroids for low n.

3.6   Matroids of rank 3

All matroids of rank 2 with $n \leq 6$ are vector matroids, and can be represented over any infinite field.

Loopless vector matroids of rank 3 are defined by line arrangements in $\mathbb{P}V \simeq \mathbb{P}^2$. So, to work with them we can draw planar pictures.

3.6.1   Decomposable matroids

The decomposable line arrangements corresponding to the connected components of ranks $3 = 1 + 1 + 1$ and $3 = 2 + 1$ are shown in the picture below.

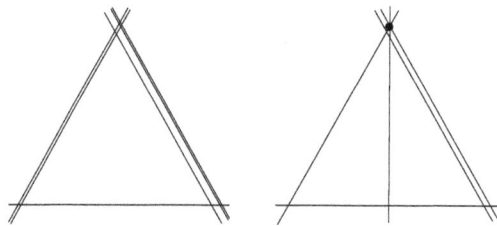

Figure 3.4: Decomposable matroids of rank 3.

In the first case, the matroids are in bijection with partitions of n into three parts. In the second case, we first need to partition $n = n_2 + n_1$ with $n_2, n_1 \geq 1$ and then partition n_2 into three or more parts. The formulas and the answers for low n are given in the following table.

	formula	4	5	6	7	8	9
3=1+1+1	$p_3(n)$	1	2	3	4	5	7
3=2+1	$p_{\geq 3}(3) + \cdots + p_{\geq 3}(n-1)$	1	3	7	14	25	42
total		2	5	10	18	30	49

Table 3.2: Number of decomposable rank-3 matroids for low n.

Since all matroids of ranks 1 and 2 are realizable over any infinite field, all decomposable rank-3 matroids are realizable over any infinite field as well.

3.6.2 Connected matroids with $n = 4$

There is only one such arrangement, four lines in general position. Up to an automorphism of $\mathbb{P}^2$, this arrangement is unique and can be given by the equations $f_1 = z_0$, $f_2 = z_1$, $f_3 = z_2$, and $f_4 = z_0 + z_1 + z_2$.

3.6.3 Connected matroids with $n = 5$

One can notice that the matroids with $(n, r) = (5, 3)$ are in bijection with the matroids with $(n, r) = (5, 2)$. This is a special case of duality, explained in Section 3.9.

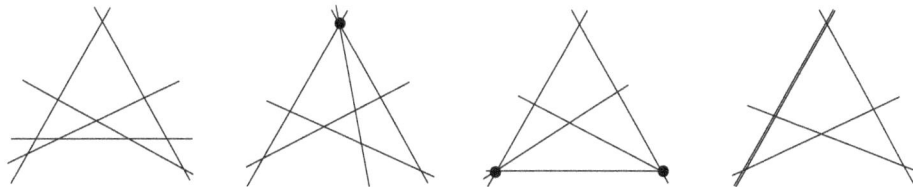

Figure 3.5: Connected matroids of rank 3 with $n = 5$.

3.6.4 Connected matroids with $n = 6$

The loopless connected matroids with $n = 6$ are all realizable. They are given in Figure 3.6.

3.6.5 Matroids with $n \geq 7$

It gets harder and harder to draw pictures and list all the possible cases as n increases. Also, as we explained in Section 3.2, for $n \geq 7$ non-realizable matroids appear, which do not correspond to any line arrangements.

On the other hand, clearly it is an algorithmically feasible problem to list all sets $\mathcal{B}$ of subsets of $\bar{n}$ satisfying Definition 3.2.2. The computation probably cannot be done in polynomial time, so it becomes unwieldy for large n.

This problem was considered in [45], where an efficient algorithm was developed and all matroids for $r = 3$, $n \leq 12$ and $r = 4$, $n \leq 10$ were enumerated. An online database maintained by the authors is available at

http://www-imai.is.s.u-tokyo.ac.jp/~ymatsu/matroid/index.html.

The first two lines in Table 3.3 are taken directly from that database.

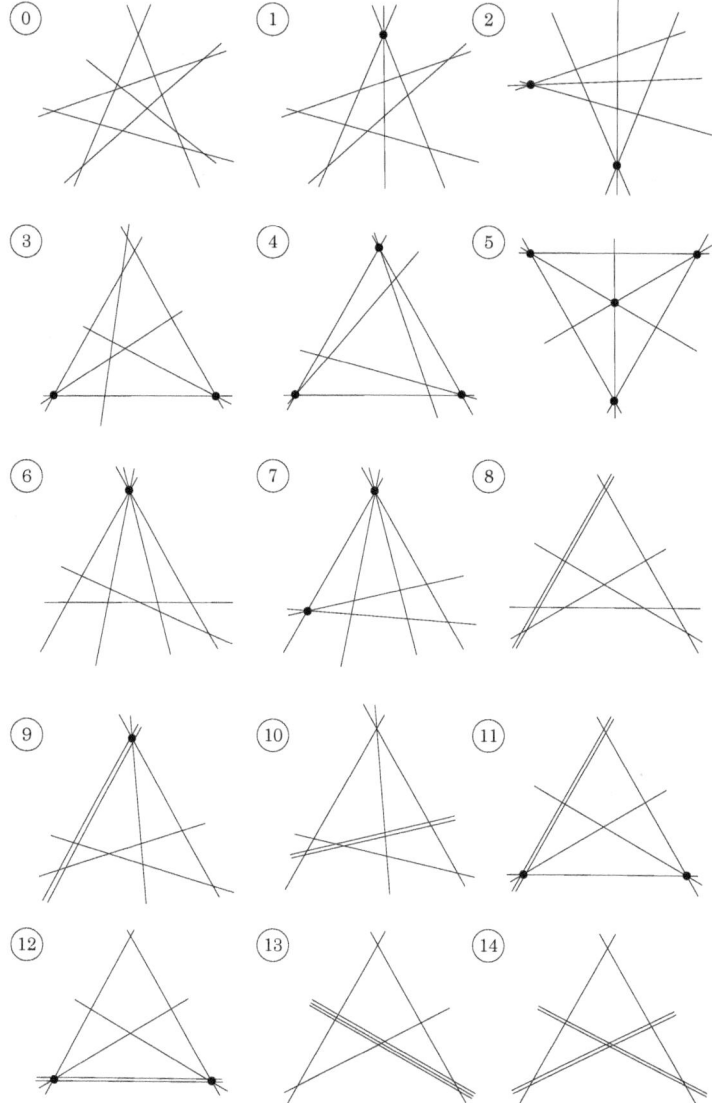

Figure 3.6: Connected matroids of rank 3 with $n = 6$.

A matroid is called *simple* if every dependent set has at least 3 elements. In other words, it is loopless and does not have parallel elements. For a realizable matroid, this means that the n hyperplanes are distinct. Clearly, one can list all connected matroids on $\overline{n}$ by considering the simple connected matroids on $\overline{m}$ for $m \le n$ and then adding multiples.

n	4	5	6	7	8	9	10	11	12
all matroids	4	13	38	108	325	1,275	10,037	298,491	31,899,134
simple conn.	1	3	8	22	67	382	5,248	232,927	28,872,971
connected	1	4	15	52	187	901			

Table 3.3: Number of rank-3 matroids for $4 \le n \le 12$.

For the connected matroids with $n = 6$, this gives $15 = 2 + 5 + 8$, where 2 comes from the simple connected matroid with $n = 4$, 5 from $n = 5$, and 8 from $n = 6$. For the connected matroids with $n = 7$, one gets $52 = 3 + 12 + 15 + 22$.

3.7 Flats

A *flat of a hyperplane arrangement* $(\mathbb{P}V, B_1, \ldots, B_n)$ is a linear subspace $Z \subset \mathbb{P}V$ of the form $Z = B(J)$ for some $J \subset \overline{n}$. This includes the empty space. For each flat Z, let

$$I(Z) = \{i \in \overline{n} \mid B_i \supset Z\} \subset \overline{n}.$$

The sets of this form are the *flats of the matroid* M. Thus, the flats of M are in bijection with the distinct sets Z. One has $Z(\varnothing) = \overline{n}$ and $Z(\mathbb{P}V) = \varnothing$. Also, for each hyperplane B_i, $I(B_i) = \{j \mid B_j = B_i\}$.

For an abstract matroid, a flat is a subset $I \subset E$ such that for any $j \notin I$ one has $r(I \cup j) > r(I)$. Another name for flats is *closed sets*.

3.8 Restrictions and contractions

There are two basic operations for a hyperplane arrangement: we can consider only a subset of the hyperplanes, and one can restrict hyperplanes to a flat. For matroids, these two operations are called *restriction* and *contraction*.

Let M be a loopless matroid corresponding to a hyperplane arrangement $(\mathbb{P}V, B_1, \ldots, B_n)$. Let $I \subset \overline{n}$ be any subset.

Definition 3.8.1. The *restriction* $M|_I$ is defined by the hyperplanes B_i, $i \in I$. Since the intersection $B(I)$ is not necessarily empty, more properly it defines a hyperplane arrangement on $\mathbb{P}V'$, where $V' = V/\{f_i = 0, \ i \in I\}$.

Definition 3.8.2. The *contraction* M/I is defined by the hyperplanes $B_j \cap B(I)$ for all $j \in I^c = \overline{n} \setminus I$. It is a true hyperplane arrangement on the projective space $B(I)$ exactly when I is a flat, so that the restrictions of linear functionals f_j to the vector space $\{f_i = 0, \ i \in I\}$ are nonzero.

For an abstract matroid M, the restriction $M|_I$ is a matroid on the set I with the rank function $r|_I$, and the contraction M/I is a matroid on I^c with the rank function $r'(A) = r(A \cup I) - r(I)$.

Note that for hyperplane arrangements it would be more intuitive to call the second operation M/I restriction, but the names come from the picture of vectors in the dual vector space V^*.

3.9   Dual matroids

Let M be a matroid on the set $E = \bar{n}$. The *dual matroid* M^* is the matroid on the same set E whose bases are the complements of the bases of M. (It takes a little work to show that the axioms of Definition 3.2.2 are satisfied.) Thus, the rank of the dual matroid is $r^* = n - r$.

The dual of a vector matroid is also a vector matroid defined as follows. Represent the vectors $f_1, \ldots, f_n \in W$ by a surjective homomorphism $k^n \to W$ and write the corresponding short exact sequence, with $\dim W = r$, $\dim W' = r^*$:

$$0 \longrightarrow W' \longrightarrow k^n \longrightarrow W \longrightarrow 0.$$

The matroid M^* is defined by dualizing this sequence to obtain

$$0 \longrightarrow W^* \longrightarrow k^n \longrightarrow (W')^* \longrightarrow 0.$$

Thus, it is represented by n vectors in the r^*-dimensional vector space $(W')^*$. One computes the rank function of M^* to be

$$r^*(I^*) = r(I) + |I^*| - r,$$

for any $I \subset \bar{n}$, $I^* = E \smallsetminus I$.

Remark 3.9.1. One has to be careful to note that a complement of a loopless matroid may have loops (zero vectors). Thus, when M corresponds to a hyperplane arrangement, the dual may not. An example is a line arrangement in $\mathbb{P}^2$ with 5 lines such that B_1 is general and $B_2, \ldots, B_5$ are distinct lines passing through a common point. The dual matroid has rank 2 and one has $r(1) = r(2345) + 1 - 3 = 0$. Thus, $f_1^* = 0$, so M^* does not correspond to a point arrangement on $\mathbb{P}^1$.

3.10   Regular matroids and degenerations of abelian varieties

3.10.1   Regular matroids

Let us also introduce regular matroids. They are not used for hyperplane arrangements, but they turn out to be important for degenerations of abelian varieties.

Definition 3.10.1. A matroid M is *regular* if it can be realized over a field of arbitrary characteristic.

A basic fact is that a regular matroid can be realized by columns of a totally unimodular matrix.

Definition 3.10.2. A matrix A with entries in $\mathbb{Z}$ is called *totally unimodular* if it has rank r and all $r \times r$ minors are 0, ±1.

Thus, a totally unimodular matrix defines a representable matroid over any field, and the set $\mathcal{B}$ of bases over any field is exactly the same. There are two additional crucial facts to know about regular matroids:

1. there are three basic types of matroids: graphic, cographic, and a special rank-5 matroid R_{10} on 10 elements, and

2. all other matroids are obtained from these elementary blocks by a kind of tensor product operation (Seymour's decomposition theorem, see below).

3.10.2 Dicings

Let A be an $r \times n$ matrix with entries in $\mathbb{Z}$ and with column vectors $f_1, \ldots, f_n$ generating $\mathbb{R}^n$. Consider a polyhedral decomposition of $\mathbb{R}^n$ obtained by cutting it along the $\mathbb{Z}^n$ hyperplanes $f_i(x) = n_i$, $i = 1, \ldots, n$, $n_i \in \mathbb{Z}$.

For each r-tuple $f_{i_1}, \ldots, f_{i_r}$ forming a basis of $\mathbb{R}^n$, the subdivision by the r systems of hyperplanes $i = i_1, \ldots, i_r$ consists of parallelohedra with vertices in a lattice $\Lambda \supset \mathbb{Z}^r$, and $|\Lambda/\mathbb{Z}^r|$ is the determinant of the corresponding $r \times r$ minor. Thus, all the vertices of the polyhedral subdivision belong to the original $\mathbb{Z}^r$ if and only if all the $r \times r$ minors are 0 or ±1.

Definition 3.10.3. A *dicing* is a polyhedral subdivision of $\mathbb{R}^g \supset \mathbb{Z}^g$ defined by a totally unimodular $g \times n$ matrix A with coefficients in $\mathbb{Z}$.

Let $r = \text{rank}(A)$. If $r = g$, the polyhedra are (finite) polytopes with vertices in $\mathbb{Z}^g$. If $r < g$, the polyhedra are infinite and are the preimages of polyhedra in $\mathbb{R}^g$ under a surjective homomorphism $\mathbb{Z}^g \twoheadrightarrow \mathbb{Z}^r$.

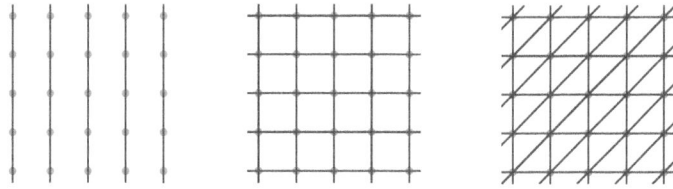

Figure 3.7: Three examples of dicings.

3.10.3   Degenerations of principally polarized abelian varieties

The limit of a 1-parameter degeneration of principally polarized abelian varieties of dimension g is described by a $\mathbb{Z}^g$-periodic polyhedral decomposition which is the preimage of a $\mathbb{Z}^r$-periodic decomposition of $\mathbb{R}^r$ into polytopes with vertices in the same lattice $\mathbb{Z}^r$. Every 1-parameter family defines a semidefinite quadratic form q on $\mathbb{Z}^g$, and the decomposition is the so-called Delaunay decomposition $\mathrm{Del}(q)$.

Dicings are special examples of Delaunay decomposition (and, in fact, the easiest ones). A dicing for the linear forms $f_1, \ldots, f_n$ is the Delaunay decomposition for a quadratic form $q = \sum_{i=1}^n a_i f_i^2$ for any $a_i > 0$. So, dicings describe a particular class of degenerations of abelian varieties.

3.10.4   Graphic matroids

Let Γ be a graph with m vertices v_i and n edges e_j. Let us pick an orientation on the edges. Then we have the chain groups and the differential map

$$\partial : C_1(\Gamma, k) = \bigoplus k e_j \quad \longrightarrow \quad C_0(\Gamma, k) = \bigoplus k v_i$$
$$e_j \quad \longmapsto \quad \mathrm{beg}(e_j) - \mathrm{end}(e_j).$$

This gives a surjection $k^n = C_1(\Gamma, k) \twoheadrightarrow \partial C_1(\Gamma, k)$.

Definition 3.10.4. The *graphic matroid* associated to the graph Γ is the vector matroid on the set $E(\Gamma)$ of edges represented by the vectors $\partial e_j \in \partial C_1(\Gamma, k)$.

Note that $\partial e_j = 0$ if and only if the edge e_j is a loop in the graph Γ. This explains the use of the term *loop* to denote linearly dependent elements in other situations.

Also, note that if the graph Γ is connected then $\partial C_1(\Gamma, k) \simeq k^{m-1}$, and this identification can be obtained by forgetting one of the vertices, i.e., erasing an arbitrary row in the $(m \times n)$ matrix with the columns ∂e_j.

3.10.5   Cographic matroids and degenerations of Jacobians

Definition 3.10.5. A *cographic matroid* is defined to be the dual to a graphic matroid.

Thus, it is a matroid on the same set $E(\Gamma)$ of edges of the graph Γ and is represented by vectors in the vector space $(\ker \partial)^* = (H_1(\Gamma, k))^* = H^1(\Gamma, k)$, the first cohomology group of Γ. In terms of cochains, we have an exact sequence

$$C^0(\Gamma, k) = \bigoplus_{i=1}^m k v_i^* \xrightarrow{d} C^1(\Gamma, k) = \bigoplus_{j=1}^n k e_j^* \longrightarrow H^1(\Gamma, k) \longrightarrow 0$$

and the n vectors are $[e_j^*] \in H^1(\Gamma, k)$.

If the graph Γ is connected, then $\dim H^1(\Gamma, k) = n - (m-1) = b_1(\Gamma)$ by Euler's formula. One has $[e_j^*] = 0$ if and only if e_j is a cut-edge (i.e., a bridge) of Γ.

Consider a one-parameter degeneration of curves C_t over a smooth curve, such that C_t is a smooth projective genus g-curve for $t \neq 0$, and C_0 is a stable curve with the dual graph Γ. Then we have a one-parameter degeneration JC_t of Jacobians, which are the easiest principally polarized abelian varieties. The limit of this family is described by the dicing for the cographic matroid of Γ.

By applying the above definition, this gives the following description of the dicing. The space $C_1(\Gamma, \mathbb{Z}) = \bigoplus_{i=1}^n \mathbb{Z} e_j$ comes with a standard Euclidean basis. We subdivide it into the standard cubes with the vertices in $C_1(\Gamma, \mathbb{Z})$ and the sides parallel to the coordinate hyperplanes. The dicing of $\mathbb{R}^g$ is obtained by intersecting these cubes with the linear subspace $H_1(\Gamma, \mathbb{R}) \subset C_1(\Gamma, \mathbb{R})$ and then pulling back to $\mathbb{R}^g$ under a surjection $\mathbb{Z}^g \twoheadrightarrow H_1(\Gamma, \mathbb{Z})$.

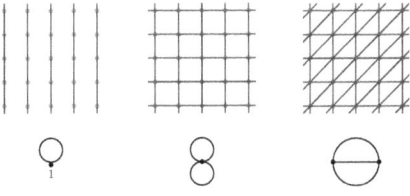

Figure 3.8: Stable graphs of curves of genus 2 and their dicings.

3.10.6 Matroid R_{10} and degenerations of Prym varieties

Definition 3.10.6. R_{10} is an exceptional matroid of rank 5 on 10 elements represented by the columns of the following matrix:

$$
\begin{pmatrix}
1 & 0 & 0 & 0 & 0 & 1 & 0 & 0 & 1 & 1 \\
0 & 1 & 0 & 0 & 0 & 1 & 1 & 0 & 0 & 1 \\
0 & 0 & 1 & 0 & 0 & 0 & 1 & 1 & 0 & 1 \\
0 & 0 & 0 & 1 & 0 & 0 & 0 & 1 & 1 & 1 \\
0 & 0 & 0 & 0 & 1 & 1 & 1 & 1 & 1 & 1
\end{pmatrix}.
$$

It is neither graphic nor cographic.

Gwena [25] gave an example of degenerations of Prym varieties, the intermediate Jacobians of cubic 3-folds, which is described by a dicing for the matroid R_{10}. In particular, these Prym varieties are not Jacobians. This implies that a generic cubic 3-fold is not rational, which is a weak form of a celebrated theorem by Clemens and Griffiths.

3.10.7   Seymour's decomposition theorem

Definition 3.10.7. Let M_1 and M_2 be matroids on the sets $S_1, S_2 \subset S'$. Define a new matroid on the symmetric difference $S = S_1 \vartriangle S_2$ by declaring its cycles (disjoint unions of circuits, i.e., minimal dependent sets) to be the symmetric differences of cycles of M_1 and M_2.

Then M is called *1-sum* if $S_1 \cap S_2 = \varnothing$, *2-sum* if $|S_1 \cap S_2| = 1$, and *3-sum* if $S_1 \cap S_2$ is a common three-point circuit.

These operations can be easily translated into operations on totally unimodular matrices.

Theorem 3.10.8 ([54]). *Any regular matroid is obtained from several graphic matroids, cographic matroids, and R_{10} by applying 1-, 2-, and 3-sum operations.*

3.10.8   Extended Torelli map

The moduli space A_g of principally polarized abelian varieties has infinitely many toroidal compactifications. Each toroidal compactification $\overline{A}_g^\tau$ corresponds to a fan τ on the space $\mathbb{R}^{g(g+1)/2}$ of quadratic forms on $\mathbb{R}^g$ such that

1. the support of τ is the cone $\overline{C}^{\mathrm{rat}}$ generated by the positive semi-definite quadratic forms with coefficients in $\mathbb{Q}$;

2. the fan τ is equivariant with respect to the action of the group $\mathrm{GL}(g, \mathbb{Z})$: cones go to cones;

3. modulo the $\mathrm{GL}(g, \mathbb{Z})$-action, there are only finitely many cones.

One particular choice for τ is the second Voronoi fan τ^{vor} defined as follows: q_1, q_2 lie in the same cone if and only if the Delaunay decompositions are the same, i.e., $\mathrm{Del}(q_1) = \mathrm{Del}(q_2)$. According to the Namikawa–Mumford theorem, the Torelli map $\mathrm{M}_g \to \mathrm{A}_g$, $C \mapsto JC$, extends to a morphism $\overline{\mathrm{M}}_g \to \overline{\mathrm{A}}_g^{\mathrm{vor}}$ from the Deligne–Mumford compactification.

Alexeev and Brunyate showed in [10] that the Torelli map also extends to a morphism $\mathrm{M}_g \to \mathrm{A}_g^{\mathrm{perf}}$ to another interesting compactification, for the perfect fan τ^{perf}. Extending this result, Melo and Viviani [46] proved that the maximal open subset U which is shared by $\mathrm{A}_g^{\mathrm{vor}}$ and $\mathrm{A}_g^{\mathrm{perf}}$ (which are birationally isomorphic, as they both contain A_g) is precisely the locus of dicings corresponding to all regular matroids of rank up to g. The compactified Torelli map factors through U.

Chapter 4

Matroid Polytopes and Tilings

Some of the results we explain here are contained in [22, 24]. Another good source on matroid polytopes is [53].

We will work in a Euclidean space $\mathbb{R}^n$ with a standard basis $e_1, \dots, e_n$. For any subset $I \subset \overline{n} = \{1, \dots, n\}$, e_I will denote the vector $\sum_{i \in I} e_i$. We will also use two special vectors, $\mathbf{0} = e_\varnothing = (0, \dots, 0)$ and $\mathbf{1} = e_{\overline{n}} = (1, \dots, 1)$.

For two vectors $x, y \in \mathbb{R}^n$, we say that $x \leq y$ if $x_i \leq y_i$ for all i.

A *polytope* is a convex hull of finitely many points in $\mathbb{R}^n$. Dually, it can be defined as a *bounded* intersection of finitely many half-spaces, i.e., by finitely many linear inequalities $\ell_s \geq 0$. (More generally, a *polyhedron* is a possibly unbounded locally finite intersection of half-spaces.) A polytope is *integral* if its vertices lie in $\mathbb{Z}^n$.

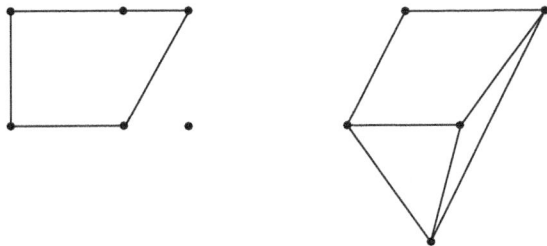

Figure 4.1: Some lattice polytopes.

Matroid polytopes form a very special class among all integral polytopes. They are defined by unusually simple inequalities. To describe them, we will denote by $x(I) = x_I$ the sum $\sum_{i \in I} x_i$, for any subset $I \subset \overline{n}$.

4.1   Base polytope and independent set polytope

4.1.1   First properties and dimension

Let M be a matroid on the set $\overline{n}$ with the independent sets $\mathcal{I} \subset 2^{\overline{n}}$ and bases $\mathcal{B} \subset 2^{\overline{n}}$.

Definition 4.1.1. The *independent set polytope* ISP_M is the convex hull of the points e_I for $I \in \mathcal{I}$. The *base polytope* BP_M is the convex hull of the points e_I with $I \in \mathcal{B}$.

In the algebraic geometry literature, mostly the second type of polytope appears, and most algebraic geometers call it *matroid polytope*. It turns out, however, that both polytopes have important applications in algebraic geometry.

Let M be a loopless vector matroid represented by vectors $f_1, \ldots, f_n$. Recall that an empty set is always independent. Therefore, $\mathbf{0}$ is always one of the vertices of ISP_M. Since $f_i \neq 0$, the points e_i also belong to ISP_M. Thus, $\dim \mathrm{ISP}_M = n$, the maximal possible.

The base polytope BP_M lies in the affine subspace $\sum n_i = r$, so it has dimension at most $n - 1$. But it can easily be lower than that.

Example 4.1.2. Let $f_1, \ldots, f_r$ be a basis of a vector space W, so that $n = r$. The independent set polytope is the cube with vertices e_I for all $I \subset \overline{n}$. However, there is only one basis, so the base polytope is the point $\mathbf{1}$ and it has dimension 0.

It is a fact that the base polytope of a connected matroid has maximal possible dimension $n - 1$. In general, suppose that M decomposes into c connected components, say $\overline{n} = S_1 \sqcup \cdots \sqcup S_c$, with rank $M_p = r_p$ and $r = \sum r_p$, $n = \sum n_p$. Then the base sets of M are of the form $I_1 \sqcup \cdots \sqcup I_c$, where $I_p \subset S_p$ is a base set. Therefore, the base polytope is the Cartesian product $\mathrm{BP}_M = \mathrm{BP}_{M_1} \times \cdots \times \mathrm{BP}_{M_c}$. Since the polytopes BP_p have dimensions $n_p - 1$, we get $\dim \mathrm{BP}_M = n - c$.

Example 4.1.3. In Figure 4.2 we can see the hyperplane arrangements for $r = 2$, $n = 4$, and their base polytopes.

Unfortunately, this is pretty much the last example which is easy to draw. For $n \geq 5$, the base polytopes of connected matroids have dimension ≥ 4, and so visualizing them becomes pretty tricky. Instead, we have to rely on other ways, for example on understanding the defining inequalities.

Definition 4.1.4. The base polytope of a generic hyperplane arrangement in $\mathbb{P}^{r-1}$ with n hyperplanes is

$$\Delta(r, n) = \mathrm{Conv}\left(e_I \mid I \subset \overline{n}, \ |I| = r \right) = \{ x \in \mathbb{R}^n \mid \mathbf{0} \leq x \leq \mathbf{1}, \ x(\overline{n}) = r \}.$$

It is called the *hypersimplex*. The matroid is the uniform matroid.

The usual simplex appears as the special case $\sigma_n = \Delta(1, n)$. The hypersimplex $\Delta(r, n)$ can be obtained from a simplex $r\sigma_n$ with sides of size r by taking the

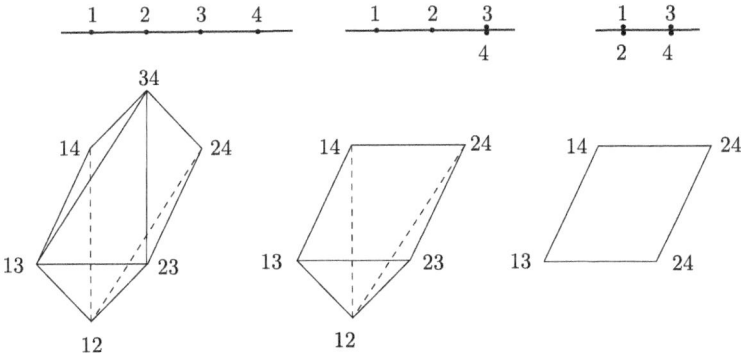

Figure 4.2: Hyperplane arrangements for $r = 2$, $n = 4$, and their base polytopes.

convex hull of the centers of the faces with r vertices (these faces are simplices $r\sigma_r$ of dimension $r - 1$).

The hypersimplex $\Delta(r, n)$ has $\binom{n}{r}$ vertices which are in bijection with the Plücker coordinates p_I. Indeed, $\Delta(r, n)$ is the moment polytope of the Grassmannian in its Plücker embedding $G(r, n) \subset \mathbb{P}^N$.

If $r \geq 2$, then $\Delta(r, n)$ has $2n$ facets (codimension 1 faces) given by the equations $x_i = 0$ and $x_i = 1$. For $r = 1$, the n faces $x_i = 1$ degenerate and become points.

4.1.2   Characterization of matroids by base polytopes

Recall that abstract matroids can be equivalently defined in terms of independent sets, base sets, rank functions, circuits, etc. Here, we get yet another definition: matroids can be defined by their base polytopes.

Going from this definition to others and back is easy, since the vertices of BP_M are the characteristic vectors of the bases of M. The following theorem of Gelfand and Serganova says precisely which polytopes appear in this way.

Theorem 4.1.5 ([24]). *A polytope in the linear space $\{x(\overline{n}) = r\} \subset \mathbb{R}^n$ with vertices of the form e_I for some $I \subset \overline{n}$ is a base polytope of some matroid of rank r if and only if all of its edges are parallel to $e_i - e_j$ for some i, j.*

The proof proceeds by observing that this property is equivalent to the base exchange property in Definition 3.2.2.

There does not appear to be an easy way to characterize the base polytopes of *representable* matroids. Of course, this question is equivalent to characterizing the representable matroids among all matroids.

4.1.3   Base polytope as a moment polytope

Let $[V] \in G(r,n)$ be a point. The closure of the T-orbit $T.[V]$ is a possibly non-normal variety in $G(r,n)$. Let us call it X' and let $f: X \to X'$ be its normalization, so that X is a toric variety in our definition. The Plücker embedding $G(r,n) \subset \mathbb{P}^N$, $N = \binom{n}{r} - 1$ is T-invariant, and the sheaf $\mathcal{O}(1)$ is T-linearized, with $\mathrm{wt}(p_I) = e_I$.

Pulling back $L = f^*\mathcal{O}(1)$, one obtains a polarized toric variety (X, L). By the correspondence, the corresponding polytope is the convex hull of the weights m such that $H^0(X, L)_m \neq 0$. A small argument shows that these weights m correspond to the Plücker coordinates p_I with $p_I(V) \neq 0$. In other words, the weights are the vectors e_I for all I, $|I| = r$, for which f_i, $i \in I$, form a basis. Thus, P is the base polytope as defined in 4.1.1 below.

Moreover, one has the following result.

Theorem 4.1.6 ([62]). *Any base polytope is totally generating (see Definition 2.3.4).*

This implies that the morphism $X \to \mathbb{P}^N$ is a closed embedding, so that $X = X'$. In other words, the subvariety $\overline{T.[V]}$ is, in fact, a normal toric variety.

When working over $\mathbb{C}$, BP_V is the moment polytope of the pair (X, L).

4.2   Facets and faces

We state the description of the faces and facets of base polytopes. The interested reader may consult [24, 53] for the proofs.

From the definition, we know the vertices of BP_M. They correspond to the bases of the matroid. For a hyperplane arrangement $(\mathbb{P}V, B_1, \ldots, B_n)$, these are the subsets $I \subset \overline{n}$ of cardinality r such that $B(I) = \bigcap_{i \in I} B_i = \varnothing$. So one can list them explicitly, but there are way too many of them. A much more economical way of describing a base polytope is by listing its facets, i.e., the minimal set of defining inequalities.

Definition 4.2.1. A flat $I \subset \overline{n}$ is *nondegenerate* if the matroids $M|_I$ and M/I are both connected.

Recall that for a hyperplane arrangement, flats are in bijection with linear subsets $Z \subset \mathbb{P}V$ of the form $Z = B(I)$ for some $I \subset \overline{n}$. In this case,

1. $M|_I$ is connected if and only if Z is *not* a transversal intersection $Z_1 \pitchfork Z_2$ of larger flats, and

2. M/I is connected if and only if the hyperplane arrangement $\{B_j \cap Z\}$ on Z is *not* a join of several smaller hyperplane arrangements. Equivalently, M/I is connected if the automorphism group $\mathrm{Aut}(Z, B_j \cap Z)$ is trivial.

Theorem 4.2.2. *The minimal set of inequalities for the base polytope BP_M is*

$$\boldsymbol{x}(\overline{n}) = r, \quad x_i \geq 0 \ \text{for } i = 1, \ldots, n, \quad \text{and} \quad \boldsymbol{x}(I) \leq r(I)$$

for all nondegenerate flats $I \neq \emptyset, \bar{n}$. For a hyperplane arrangement, the latter means flats $Z \neq \mathbb{P}V, \emptyset$.

The inequality $\boldsymbol{x}(I) \leq r(I)$ holds for any subset $I \subset \bar{n}$ but these inequalities are redundant unless I is a nondegenerate flat.

Definition 4.2.3. The inequalities $\boldsymbol{x}(\bar{n}) = r$ and $x_i \geq 0$ are always present. So usually we take them for granted. We call the rest of the inequalities $\boldsymbol{x}(I) \leq r(I)$ *essential.*

Example 4.2.4. For a generic hyperplane arrangement, the only nondegenerate flats are 1-element sets $\{i\}$, since all the intersections are transversal. So, the essential inequalities are $x_i \leq 1$.

Example 4.2.5. For the following line arrangement of six lines in $\mathbb{P}^2$ the essential inequalities are

$$x_1 \leq 1, \quad x_{23} \leq 1, \quad x_4 \leq 1, \quad x_5 \leq 1, \quad x_{1236} \leq 2, \quad x_{456} \leq 2.$$

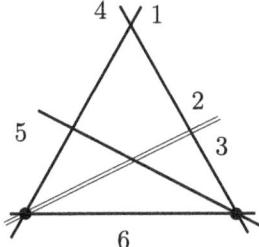

Figure 4.3: A line arrangement with six lines.

Thus, there are four inequalities for four out of five distinct lines in the picture, and two inequalities for the points where three of them intersect. The flat $\{6\}$ is degenerate, since there are only two special points on the line $Z = B_6$. The corresponding inequality $x_6 \leq 1$ follows from $x_{1236} \leq 2$, $x_{456} \leq 2$, and $x_{123456} = 3$.

4.3 Matroid polytopes and log canonical singularities

Recall the definition of lc singularities from Section 1.2, and let us look at Figure 4.3 above. For which coefficients b_i is the pair $(\mathbb{P}^2, \sum b_i B_i)$ log canonical ? To be lc along the line Z_{23}, one must have $b_{23} \leq 1$. To be lc at the intersection points of three or more lines, one must have $b_{1236} \leq 2$ and $b_{456} \leq 2$. And, of course, there are the inequalities $b_i \leq 1$ which are usually included in the definition of lc singularities to begin with.

Thus, we see that, at least in this example, the inequalities for lc singularities are exactly the same as the defining inequalities for the base polytope. In fact, this is true in general. We state without proof several theorems from [7]. Recall the following definition.

Definition 4.3.1. A pair $(X, \sum B = b_i B_i)$ is called *log Calabi–Yau* if $K_X + B \sim_{\mathbb{Q}} 0$; it is called *log Fano* if $-(K_X + B)$ is ample; and it is called *of general type* if $K_X + B$ is big (e.g., ample).

For a hyperplane arrangement, $K_{\mathbb{P}^{r-1}} \sim -rH$ and $B_i \sim H$, where H is the class of a hyperplane. Therefore, the pair $(\mathbb{P}^{r-1}, \sum b_i B_i)$ is log Calabi–Yau (resp. log Fano, or of general type) if and only if $\sum b_i = r$ (resp. $\sum b_i < r$, or $\sum b_i > r$).

Theorem 4.3.2. *A log Calabi–Yau hyperplane arrangement* $(\mathbb{P}V, \sum b_i B_i)$ *is* lc *if and only if* $\boldsymbol{b} \in \mathrm{BP}_M$.

Thus, the inequalities for the base polytope appear most naturally in the case of log Calabi–Yau hyperplane arrangements. The independent set polytope is best suited to log Fano pairs:

Theorem 4.3.3. *A log Fano hyperplane arrangement is* lc *if and only if* $\boldsymbol{b} \in \mathrm{ISP}_M$.

The independent set polytope also appears in the condition for an arbitrary hyperplane arrangement to be lc at a given point. To formulate this, we need another piece of notation.

Definition 4.3.4. Let $(\mathbb{P}V, B_1, \dots, B_n)$ be a hyperplane arrangement. For a point $p \in \mathbb{P}V$, we denote by $I(p)$ the set of $i \in \overline{n}$ such that $p \in B_i$. For a vector $\boldsymbol{b} \in \mathbb{R}^n$, we denote by $\boldsymbol{b}|_{I(p)}$ the vector $\boldsymbol{x}$ with $x_i = b_i$ if $i \in I(p)$ and $x_i = 0$ otherwise.

Theorem 4.3.5. *A hyperplane arrangement* $(\mathbb{P}V, \sum b_i B_i)$ *is* lc *at a point* $p \in \mathbb{P}V$ *if and only if* $\boldsymbol{b}|_{I(p)} \in \mathrm{ISP}_M$.

4.4   Cuts of polytopes and log canonical singularities

In these lectures, we are interested in the stable pairs, which are of general type, not Calabi–Yau nor Fano. Below, we describe how the combinatorics of matroid polytopes has to be adjusted to handle this case. Again, the proofs can be found in [7].

Definition 4.4.1. We define the $\boldsymbol{b}$-*cut of the hypersimplex* by

$$\Delta_{\boldsymbol{b}}(r, n) = \left\{ \boldsymbol{x} \in \mathbb{R}^n \mid \boldsymbol{0} \le \boldsymbol{x} \le \boldsymbol{b}, \ \boldsymbol{x}(\overline{n}) = r \right\}.$$

Theorem 4.4.2. *A hyperplane arrangement* $(\mathbb{P}V, \sum b_i B_i)$ *of general or log Calabi–Yau type is* lc *if and only if* $\Delta_{\boldsymbol{b}} \subset \mathrm{BP}_M$.

Example 4.4.3. $\Delta_{\boldsymbol{1}}(r, n) = \Delta(r, n)$, the ordinary hypersimplex. Thus, $\Delta_{\boldsymbol{1}} \subset \mathrm{BP}_M$ if and only if $\mathrm{BP}_M = \Delta(r, n)$. The only hyperplane arrangement with this base polytope is the generic hyperplane arrangement. So, the pair $(\mathbb{P}V, \sum B_i)$ is lc only for generic hyperplane arrangements.

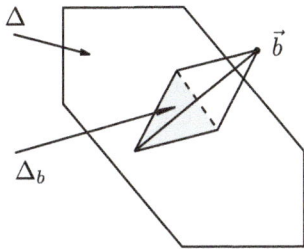

Figure 4.4: b-cut hypersimplex.

Example 4.4.4. Let $b > 1$, i.e., all $b_i \geq 1$ with at least one inequality strict. Then Δ_b is strictly bigger than $\Delta(r, n)$, so it is not contained in *any* base polytope. Thus, $(\mathbb{P}V, \sum B_i)$ is not lc, which is clear since there is some coefficient $b_i > 1$.

Example 4.4.5. If b is a vector with $b(\overline{n}) = r$, then the polytope $\Delta_b(r, n)$ is a single point $\{b\}$. So, when the pair $(\mathbb{P}V, \sum b_i B_i)$ is of log Calabi–Yau type, the condition $\Delta_b \subset \mathrm{BP}_M$ is the same as the condition $b \in \mathrm{BP}_M$.

Next, we address the question of when the hyperplane arrangement is lc at a given point.

Definition 4.4.6. For $p \in \mathbb{P}V$, define Δ_b^p to be the (possibly empty) face of Δ_b, where $x_i = b_i$ for all $i \in I(p)$.

Theorem 4.4.7. *Let $(\mathbb{P}V, \sum b_i B_i)$ be a hyperplane arrangement of general type. Suppose that $\mathrm{BP}_M \cap \Delta_b \neq \varnothing$. Then $(\mathbb{P}V, \sum b_i B_i)$ is lc at p if and only if*

$$\mathrm{BP}_M \cap \Delta_b^p \neq \varnothing.$$

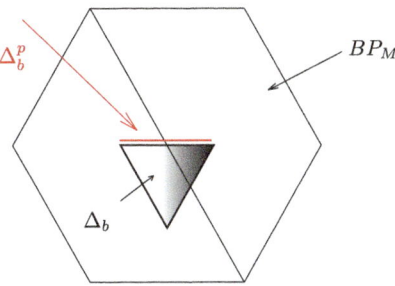

Figure 4.5: BP_M and Δ_b^p.

4.5   Matroid tilings

Definition 4.5.1. A *tiling* is a collection of polytopes Q_j in $\mathbb{R}^n$ which is *face-fitting*: the intersection of any two of them, $Q_{j_1} \cap Q_{j_2}$ is either empty or is a face of both.

Definition 4.5.2. A *partial matroid tiling* is a tiling consisting of base polytopes in the hypersimplex $\Delta(r,n) \smallsetminus \bigcup_{i=1}^{n}\{x_i = 0\}$. It does not have to cover $\Delta(r,n)$ completely.

We ignore the base polytopes contained in one of the spaces $\{x_i = 0\}$, i.e., base polytopes of matroids that contain loops $f_i = 0$, since those do not correspond to hyperplane arrangements.

Definition 4.5.3. A tiling of the *b*-cut hypersimplex Δ_b is a partial matroid tiling such that $\bigcup \mathrm{BP}_{M_j} \supset \Delta_b$ and such that all base polytopes BP_{M_j} intersect Δ_b.

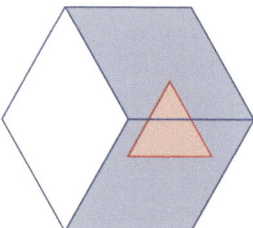

Figure 4.6: A matroid tiling of the *b*-cut hypersimplex Δ_b.

The intuition for the algebro-geometric meaning of such tilings is as follows. When a single base polytope BP_M covers Δ_b, the hyperplane arrangement $(\mathbb{P}V, \sum b_i B_i)$ is lc and gives a point in the moduli space of stable pairs. But if $\Delta_b \not\subset \mathrm{BP}_M$, then several base polytopes BP_{M_j} are needed to cover it. In this case, the projective space $\mathbb{P}^{r-1}$ degenerates, and the stable pair $X = \bigcup X_j$ is non-normal and consists of several irreducible components X_j. The irreducible components X_j thus correspond to several non-lc hyperplane arrangements $(\mathbb{P}V_j, \sum B_i^{(j)})$. One should think of them as the "complementary" degenerations of $\mathbb{P}^{r-1}$ with n hyperplanes. They complement each other to give the entire stable pair (X, B).

4.6   Rank-2 case

The combinatorics of matroid tilings in rank 2 is exactly the same as the combinatorics of stable weighted graphs. Vertices of a graph correspond to maximal-dimensional polytopes and edges correspond to facets. For the case *b* = 1, this theory was established in [33]. The proof given below for the general case is very similar.

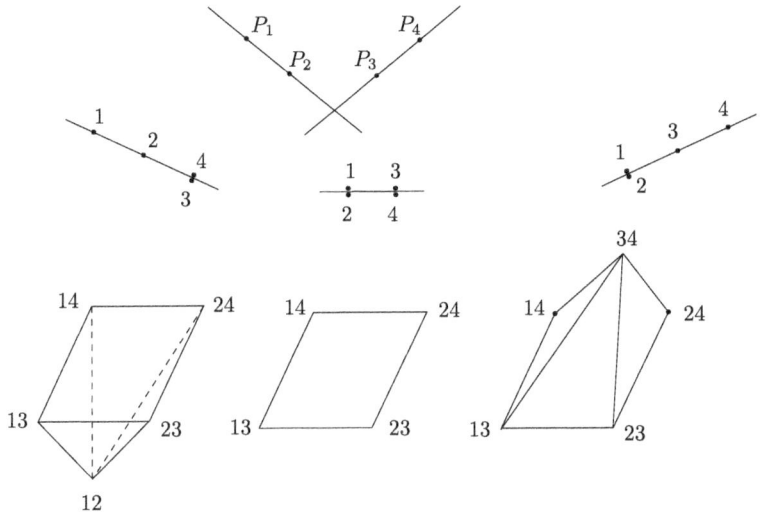

Figure 4.7: Matroid tiling corresponding to a stable curve.

Example 4.6.1. A stable curve with $n = 4$ points, its dual graph, the corresponding matroid tiling and hyperplane arrangements are shown in Figure 4.7.

Recall from 3.5 that a hyperplane arrangement in rank 2 is the same as n points on $\mathbb{P}^1$ which are allowed to coincide, but there should be at least $p \geq 2$ distinct points. The base polytopes are maximal-dimensional if $p \geq 3$ or have codimension 1 if $p = 2$ (in which case the polytope is a product of two simplices $\sigma_J \times \sigma_{J^c}$).

Let $\bigcup_j \mathrm{BP}_{M_j}$ be a partial matroid tiling of $\Delta(2, n)$. We will associate to it a graph Γ, as follows. It will be convenient to work with half-edges, or "flags".

1. To each maximal-dimensional base polytope we associate a vertex of Γ. The essential inequalities of BP_M are $x(J_s) \leq 1$ for a partition $\overline{n} = J_1 \sqcup \cdots \sqcup J_p$ with $p \geq 3$.

2. Further, to this vertex we add p half-edges going away from it, one for each of the sets J_s.

3. To each codimension 1 polytope $\mathrm{BP}_M = \{x(J) = x(J^c) = 1\}$ which is a common facet of two maximal-dimensional base polytopes with essential inequalities $x(J) \leq 1$ and $x(J^c) \leq 1$, we associate an internal edge of Γ which consists of two half-edges J and J^c.

4. To each codimension 1 polytope which has only one maximal-dimensional neighbor with essential inequality $x(J) \leq 1$ we associate an end of Γ and we mark it J.

Thus, all internal vertices of Γ have valency at least three, plus the existence of a certain number of ends.

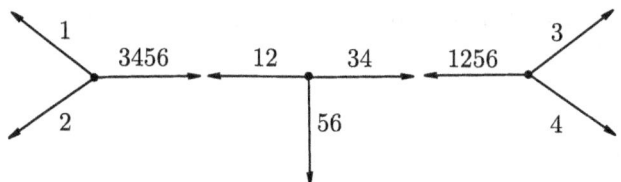

Figure 4.8: Graph Γ describing a partial matroid tiling of $\Delta(2, n)$.

Theorem 4.6.2. *Let* $\bigcup_j \mathrm{BP}_{M_j}$ *be a partial tiling of* $\Delta(2, n)$ *by base polytopes which is maximal-dimensional, and connected in codimension* 1. *Then the graph* Γ *is a tree, and the ends of* Γ *correspond to the parts in a partition* $K_1 \sqcup \cdots \sqcup K_m$ *of* $\overline{n}$ *into* $m \geq 3$ *parts. Also, the set* $Q = \bigcup_j \mathrm{BP}_{M_j}$ *is a convex polytope.*

Proof. For each edge e, the polytope BP_e is an intersection of $\Delta(2, n)$ with the hyperplane $\boldsymbol{x}(J) = \boldsymbol{x}(J^c) = 1$. Thus, it cuts the hypersimplex into two disjoint parts, and removing it disconnects the tiling. Therefore, every edge e is a bridge, so Γ is a forest. Since Q is connected in codimension 1, the graph Γ is connected, so it is a tree.

The outside boundary of Q consists of the facets $x_i \geq 0$ and the facets of the form $x(J) \leq 1$ for the ends of Γ. In both cases, Q lies entirely in the corresponding half-space. For the facets $x_i \geq 0$ this is true by the definition of base polytopes. For the facets $x(J) \leq 1$ this is true because the facet $x(J) = 1$ is the intersection of $\Delta(r, n)$ with a hyperplane and because the tiling is connected in codimension 1. Thus, Q is the intersection of the half-spaces given by the inequalities for the facets. Therefore, it is a polytope.

Now, start with any vertex v for a maximal-dimensional base polytope in our tiling. It has $p \geq 3$ half-edges $J_1, \ldots, J_p$. The half-edge J_1 is either an end marked by J_1, or it is half of an internal edge marked J_1, J_1^c. The half of this edge leads to a vertex corresponding to a partition $J_1^c, \ldots$, where the other parts are a partition of J_1. Thus, the set J_1 gets subdivided. If we continue this path outward away from v, it will be subdivided further, etc. Following all the paths from v outward, we eventually get to a partition $K_1 \sqcup \cdots \sqcup K_m$ of $\overline{n}$ refining $J_1 \sqcup \cdots \sqcup J_p$, in which all parts K_s correspond to the ends of the graph Γ. $\square$

4.7   Rank-3 case

Recall that we listed all hyperplane arrangements in $\mathbb{P}^2$ with $n \leq 6$ lines in 3.6. By 4.2.2, the essential inequalities of the corresponding base polytopes are in bijection with the nondegenerate flats. In $\mathbb{P}^2$, the nontrivial flats are lines and points. Therefore, we get:

1. For each line Z with three or more special points, let $I = \{i \in \overline{n} \mid B_i = Z\}$. Then, we get the inequality $x_I \leq 1$.

2. For each point Z with three or more lines passing through it, let $I = \{i \in \overline{n} \mid Z \in B_i\}$. Then, we get the inequality $x_I \leq 2$.

Below, there is a list of all base polytopes for $n \leq 3$, and all complete matroid tilings for $n \leq 6$. Further, one can check computationally that any partial tiling connected in codimension 1 for $n \leq 6$ can be extended to a complete tiling. Thus, all partial tilings in these cases are subtilings of a complete tiling.

4.7.1   The case $n = 4$

The simplex $\Delta(3,4) = \Delta(1,4) = \sigma_4$ has no subdivisions. So, it has only the trivial tiling: the single polytope $\Delta(3,4)$ itself.

4.7.2   The case $n = 5$

In Table 4.1, we list the four base polytopes corresponding to the four line arrangements in Figure 3.5.

no.	(volume) essential inequalities
0	(4) ∅
1	(3) $x_{123} \leq 2$
2	(2) $x_{125} \leq 2$, $x_{345} \leq 2$
3	(1) $x_{12} \leq 1$

Table 4.1: Base polytopes in $\Delta(3,5)$.

The meaning of the "volume" of a base polytope will be explained in ???. For now, let me just say that the volumes are integral and they add up to $(n-3)^2$ in a complete tiling of $\Delta(3,n)$.

Up to S_5, there are only two nontrivial base matroid tilings listed in Table 4.2.

no.	(volume) polytope
1	(3) $x_{123} \leq 2$ (1) $x_{45} \leq 1$
2	(2) $x_{125} \leq 2$, $x_{345} \leq 2$ (1) $x_{34} \leq 1$ (1) $x_{12} \leq 1$

Table 4.2: Matroid tilings of $\Delta(3,5)$.

4.7.3   The case $n = 6$

In Table 4.3, we list the fifteen base polytopes corresponding to the line arrangements in Figure 3.6 of Section 3.6.4.

The five polytopes marked with a star are *rigid*. Each of them corresponds to a rigid line arrangement which has no moduli: there is a unique line arrangement of this type up to isomorphism. For all the other line arrangements, there are positive-dimensional families of the same type (for example, there is a four-dimensional family of six lines in general position). The base polytope for a non-rigid arrangement can be split into several smaller base polytopes.

no.	(volume) essential inequalities
0	(9) $\varnothing$
1	(8) $x_{123} \leq 2$
2	(7) $x_{123} \leq 2$, $x_{456} \leq 2$
3	(7) $x_{123} \leq 2$, $x_{145} \leq 2$
4	(6) $x_{123} \leq 2$, $x_{345} \leq 2$, $x_{561} \leq 2$
5*	(5) $x_{356} \leq 2$, $x_{246} \leq 2$, $x_{145} \leq 2$, $x_{123} \leq 2$
6	(5) $x_{1234} \leq 2$
7	(4) $x_{156} \leq 2$, $x_{1234} \leq 2$
8	(4) $x_{12} \leq 1$
9	(3) $x_{1234} \leq 2$, $x_{12} \leq 1$
10	(3) $x_{345} \leq 2$, $x_{12} \leq 1$
11*	(2) $x_{356} \leq 2$, $x_{1234} \leq 2$, $x_{12} \leq 1$
12*	(2) $x_{1234} \leq 2$, $x_{1256} \leq 2$
13*	(1) $x_{123} \leq 1$
14*	(1) $x_{34} \leq 1$, $x_{12} \leq 1$

Table 4.3: Base polytopes in $\Delta(3,6)$.

Modulo S_6, there are twenty-five nontrivial tilings listed in Table 4.4, which were found by using a computer. Some of the base polytopes can be split into unions of smaller base polytopes. The seven cases marked with * are the *rigid* tilings, which can not be split any further.

no.	(volume) polytope			
1	(8) $x_{456} \leq 2$	(1) $x_{123} \leq 1$		
2	(7) $x_{123} \leq 2$, $x_{456} \leq 2$	(1) $x_{123} \leq 1$	(1) $x_{456} \leq 1$	
3	(7) $x_{124} \leq 2$, $x_{456} \leq 2$	(1) $x_{123} \leq 1$	(1) $x_{356} \leq 1$	

4	(6) $x_{124} \leq 2$, $x_{135} \leq 2$, $x_{456} \leq 2$ (1) $x_{123} \leq 1$ (1) $x_{246} \leq 1$ (1) $x_{356} \leq 1$
5	(5) $x_{1234} \leq 2$ (4) $x_{56} \leq 1$
6	(5) $x_{1234} \leq 2$ (3) $x_{123} \leq 2$, $x_{56} \leq 1$ (1) $x_{456} \leq 1$
7*	(5) $x_{124} \leq 2$, $x_{135} \leq 2$, $x_{236} \leq 2$, $x_{456} \leq 2$ (1) $x_{123} \leq 1$ (1) $x_{145} \leq 1$ (1) $x_{246} \leq 1$ (1) $x_{356} \leq 1$
8	(4) $x_{1234} \leq 2$, $x_{156} \leq 2$ (4) $x_{56} \leq 1$ (1) $x_{234} \leq 1$
9	(4) $x_{1234} \leq 2$, $x_{156} \leq 2$ (3) $x_{123} \leq 2$, $x_{56} \leq 1$ (1) $x_{234} \leq 1$ (1) $x_{456} \leq 1$
10	(4) $x_{1234} \leq 2$, $x_{156} \leq 2$ (3) $x_{234} \leq 2$, $x_{56} \leq 1$ (1) $x_{156} \leq 1$ (1) $x_{234} \leq 1$
11	(3) $x_{1234} \leq 2$, $x_{12} \leq 1$ (3) $x_{1256} \leq 2$, $x_{56} \leq 1$ (3) $x_{3456} \leq 2$, $x_{34} \leq 1$
12	(3) $x_{1234} \leq 2$, $x_{34} \leq 1$ (3) $x_{1256} \leq 2$, $x_{56} \leq 1$ (2) $x_{1234} \leq 2$, $x_{1256} \leq 2$ (1) $x_{34} \leq 1$, $x_{56} \leq 1$
13	(3) $x_{1234} \leq 2$, $x_{34} \leq 1$ (3) $x_{3456} \leq 2$, $x_{56} \leq 1$ (2) $x_{1234} \leq 2$, $x_{1256} \leq 2$ (1) $x_{12} \leq 1$, $x_{56} \leq 1$
14	(3) $x_{1256} \leq 2$, $x_{56} \leq 1$ (3) $x_{3456} \leq 2$, $x_{34} \leq 1$ (2) $x_{1234} \leq 2$, $x_{12} \leq 1$, $x_{356} \leq 2$ (1) $x_{124} \leq 1$
15	(3) $x_{1234} \leq 2$, $x_{12} \leq 1$ (2) $x_{1234} \leq 2$, $x_{3456} \leq 2$ (2) $x_{1256} \leq 2$, $x_{3456} \leq 2$ (1) $x_{12} \leq 1$, $x_{56} \leq 1$ (1) $x_{34} \leq 1$, $x_{56} \leq 1$
16	(3) $x_{3456} \leq 2$, $x_{56} \leq 1$ (2) $x_{1234} \leq 2$, $x_{3456} \leq 2$ (2) $x_{56} \leq 1$, $x_{3456} \leq 2$, $x_{124} \leq 2$ (1) $x_{356} \leq 1$ (1) $x_{12} \leq 1$, $x_{56} \leq 1$
17	(3) $x_{1234} \leq 2$, $x_{12} \leq 1$ (2) $x_{56} \leq 1$, $x_{1256} \leq 2$, $x_{3456} \leq 2$ (2) $x_{34} \leq 1$, $x_{3456} \leq 2$, $x_{125} \leq 2$ (1) $x_{12} \leq 1$, $x_{56} \leq 1$ (1) $x_{346} \leq 1$
18	(3) $x_{3456} \leq 2$, $x_{34} \leq 1$ (2) $x_{1256} \leq 2$, $x_{3456} \leq 2$, $x_{56} \leq 1$ (2) $x_{12} \leq 1$, $x_{3124} \leq 2$, $x_{356} \leq 2$ (1) $x_{12} \leq 1$, $x_{56} \leq 1$ (1) $x_{124} \leq 1$
19	(3) $x_{1256} \leq 2$, $x_{56} \leq 1$ (2) $x_{125} \leq 2$, $x_{3456} \leq 2$, $x_{34} \leq 1$ (2) $x_{356} \leq 2$, $x_{1234} \leq 2$, $x_{12} \leq 1$ (1) $x_{124} \leq 1$ (1) $x_{346} \leq 1$
20*	(2) $x_{1234} \leq 2$, $x_{1256} \leq 2$ (2) $x_{1234} \leq 2$, $x_{3456} \leq 2$ (2) $x_{1256} \leq 2$, $x_{3456} \leq 2$ (1) $x_{12} \leq 1$, $x_{34} \leq 1$ (1) $x_{12} \leq 1$, $x_{56} \leq 1$ (1) $x_{34} \leq 1$, $x_{56} \leq 1$
21*	(2) $x_{1256} \leq 2$, $x_{3456} \leq 2$, $x_{56} \leq 1$ (2) $x_{1234} \leq 2$, $x_{1256} \leq 2$, $x_{12} \leq 1$ (2) $x_{256} \leq 2$, $x_{1234} \leq 2$, $x_{34} \leq 1$ (1) $x_{12} \leq 1$, $x_{56} \leq 1$ (1) $x_{34} \leq 1$, $x_{56} \leq 1$ (1) $x_{134} \leq 1$
22*	(2) $x_{56} \leq 1$, $x_{1256} \leq 2$, $x_{3456} \leq 2$ (2) $x_{1256} \leq 2$, $x_{346} \leq 2$, $x_{12} \leq 1$ (2) $x_{256} \leq 2$, $x_{1234} \leq 2$, $x_{34} \leq 1$ (1) $x_{125} \leq 1$ (1) $x_{34} \leq 1$, $x_{56} \leq 1$ (1) $x_{134} \leq 1$
23*	(2) $x_{1256} \leq 2$, $x_{134} \leq 2$, $x_{56} \leq 1$ (2) $x_{12} \leq 1$, $x_{1234} \leq 2$, $_{1256} \leq 2$ (2) $x_{256} \leq 2$, $x_{1234} \leq 2$, $x_{34} \leq 1$ (1) $x_{256} \leq 1$ (1) $x_{34} \leq 1$, $x_{56} \leq 1$ (1) $x_{134} \leq 1$
24*	(2) $x_{1256} \leq 2$, $x_{134} \leq 2$, $x_{56} \leq 1$ (2) $x_{1234} \leq 2$, $x_{1256} \leq 2$ (2) $x_{156} \leq 2$, $x_{1234} \leq 2$, $x_{34} \leq 1$ (1) $x_{256} \leq 1$ (1) $x_{34} \leq 1$, $x_{56} \leq 1$ (1) $x_{234} \leq 1$
25*	(2) $x_{1234} \leq 2$, $x_{456} \leq 2$, $x_{12} \leq 1$ (2) $x_{125} \leq 2$, $x_{3456} \leq 2$, $x_{34} \leq 1$ (2) $x_{234} \leq 2$, $x_{1256} \leq 2$, $x_{56} \leq 1$ (1) $x_{346} \leq 1$ (1) $x_{123} \leq 1$ (1) $x_{156} \leq 1$

Table 4.4: Matroid tilings of $\Delta(3,6)$.

4.8   Tropical projective spaces and Dressians

A tropical projective linear subspace $\mathbb{TP}^{r-1} \subset \mathbb{TP}^{n-1}$ is a balanced polyhedral complex of dimension $r-1$ in $\mathbb{R}^{n-1}$ with a specified behavior at infinity.

The balancing condition is as follows. For each codimension one polyhedron P where k maximal-dimensional polyhedra $Q_1, \ldots, Q_k$ meet, let $\boldsymbol{m}_1, \ldots, \boldsymbol{m}_k$ be the integral normal vectors to P in Q_i. Then, one must have $\sum_{i=1}^k \boldsymbol{m}_i = 0$.

For example, a tropical $\mathbb{TP}^1 \subset \mathbb{TP}^{n-1}$ is a polyhedral complex of dimension 1. In this case, Q_i are line segments meeting at a point P, and $\boldsymbol{m}_i$ are the integral generators in the direction of $Q_i - P$. The complex should be a tree with n infinite ends going off to infinity in the directions $\boldsymbol{e}_1, \ldots, \boldsymbol{e}_n$. Here, the $\boldsymbol{e}_i$'s are n vectors in $\mathbb{Z}^n$ such that $\sum_{i=1}^n \boldsymbol{e}_i = 0$ (since $\mathbb{TP}^{n-1}$ is a *projective* space with n homogeneous coordinates).

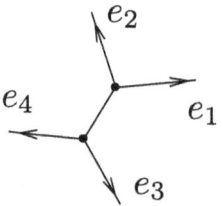

Figure 4.9: A tropical line $\mathbb{TP}^1 \subset \mathbb{TP}^3$.

Tropical projective linear subspaces $\mathbb{TP}^{r-1} \subset \mathbb{TP}^{n-1}$ are closely related to stable hyperplane arrangements with weight $\boldsymbol{b} = \boldsymbol{1}$ (the correspondence can be extended to the case of arbitrary weight $\boldsymbol{b}$ by introducing tropical projective spaces with more general behavior at infinity). Briefly, tropical $\mathbb{TP}^{r-1} \subset \mathbb{TP}^{n-1}$ correspond to *smoothings* of stable hyperplane arrangements, or *one-parameter degenerations of stable hyperplane arrangements with a smooth generic fiber*, and capture the essential combinatorial part of such smoothings. In particular, the non-smoothable stable hyperplane arrangements do not appear this way.

Let us explain this in the one-dimensional case. Consider a stable n-pointed curve $(C_0, P_1, \ldots, P_n)$ of arithmetic genus 0. Its dual graph is a tree Γ with n ends; each internal vertex has degree at least three. A tropical $\mathbb{TP}^1 \subset \mathbb{TP}^{n-1}$ as a graph is the same tree Γ with n marked ends, but it contains strictly more information than Γ. Namely, for each internal edge e there is its length $c_e > 0$ measured in the lattice units: distance between two lattice points in the chosen direction is taken to be 1.

An edge e of Γ corresponds to a singular point Q of C_0. Let $\pi : C \to S$ be a smoothing over a one-dimensional regular base with a local parameter t. Then in a neighborhood of Q the family has an equation $xy = t^n$. Thus, we may associate to e the length $c_e = n$, which is a positive integer. The tropical curve is an abstraction of this construction, with c_e allowed to be any positive real number. One can also obtain more general lengths by considering families over $S = \operatorname{Spec} R$, where R is a ring of dimension one, with a non-discrete valuation taking values in $\mathbb{Q}$ or $\mathbb{R}$. For example, one can take $R = \mathbb{C}[t^\alpha, \ \alpha \in \mathbb{R}_{>0}]$.

Similarly, *tropical planes in* $\mathbb{TP}^{n-1}$ correspond to one-parameter degenerations of the pairs $(\mathbb{P}^2, B_1, \ldots, B_n)$. They are dual metric versions of tilings of $\Delta(3, n)$, with the additional smoothing data. In terms of toric geometry, tilings live in the lattice M and a tropical linear subspace is determined by a point in a fan living in the dual space $N_{\mathbb{R}}$.

Thus, the combinatorial types of tropical planes in $\mathbb{TP}^{n-1}$ describe not all stable hyperplane arrangements, but only those that have a smoothing. These are the most important ones, anyway, and they correspond to points in the main irreducible component of the moduli space.

The combinatorial types of the tropical $\mathbb{TP}^{r-1}$ in $\mathbb{TP}^{n-1}$ are in bijection with cones in a fan that was called the *Dressian* in [32]. This is a subfan of the secondary fan of $\Delta(r, n)$. Here, recall from Chapter 2 that the secondary fan describes the smoothable stable toric varieties over the Plücker projective space $\mathbb{P}^N$, $N = \binom{n}{r} - 1$. The Dressian describes smoothings of stable toric varieties lying over a closed subset of $\mathbb{P}^N$, the Grassmannian $G(r, n)$.

The Dressians $\mathrm{Dr}(3, 6)$ and $\mathrm{Dr}(3, 7)$ were computed in [32, 56]. In particular, the seven generic tropical planes in $\mathbb{TP}^5$ from [32, Fig.1] are the same as the seven rigid types marked with a star in Table 4.4. For $n = 7$, the authors list 94 generic tropical planes in $\mathbb{TP}^6$ in http://www.uni-math.gwdg.de/jensen/Research/G3_7/grassmann3_7.html.

4.9 Dual matroid polytopes and dual tilings

Consider the linear change of coordinates $x_i^* = 1 - x_i$. This defines an involution $\iota : \mathbb{R}^n \to \mathbb{R}^n$. Under this linear transformation, the image of the hypersimplex

$$\Delta(r, n) = \left\{ 0 \le x_i \le 1, \ \sum_{i=1}^n x_i = r \right\}$$

is the hypersimplex

$$\Delta(r^*, n) = \left\{ 0 \le x_i^* \le 1, \ \sum_{i=1}^n x_i^* = n - r = r^* \right\}.$$

For any $I \subset \overline{n}$, $\iota(e_I) = e_{I^*}$, where $I^* = \overline{n} \smallsetminus I$. Since the vertices of the base polytope BP_M are the vectors e_I for the bases of M, and since the bases of the dual matroid M^* are the complementary sets I^*, we see that $\iota(\mathrm{BP}_M) = \mathrm{BP}_{M^*}$.

In terms of the inequalities, BP_M is defined by $0 \le x_i \le 1$ and $\boldsymbol{x}(I) \le r(I)$ for all $I \subset \overline{n}$ (this system is overdetermined). After the coordinate change, we get $0 \le x_i^* \le 1$ and the inequalities

$$|I| - \boldsymbol{x}^*(I) \le r(I) \iff \boldsymbol{x}^*(I) \ge |I| - r(I) \iff \boldsymbol{x}^*(I^*) \le r^* - |I| + r(I) = r^*(I^*),$$

which are precisely the defining inequalities of the base polytope BP_{M^*}.

The image of a $\boldsymbol{b}$-cut hypersimplex $\Delta(r, n)_{\boldsymbol{b}}$ is the polytope

$$\left\{ 1 - b_i \leq x_i^* \leq 1, \ \sum x_i = r^* \right\}.$$

This is a $\boldsymbol{b}^*$-cut hypersimplex only in two cases, namely when $\boldsymbol{b} = \mathbf{1}$, or when $\Delta_{\boldsymbol{b}}$ is a point (i.e., $\boldsymbol{b}(\overline{n}) = r$ and $\boldsymbol{b}^* = \mathbf{1} - \boldsymbol{b}$). Thus, the dual of a matroid tiling of $\Delta(r, n)$ is a matroid tiling of $\Delta(r^*, n)$, but for the $\boldsymbol{b}$-cuts this generally does not work.

4.10   Mnev's universality theorem

The line arrangements in $\mathbb{P}^2$ with a fixed base polytope $P = \mathrm{BP}_M$ form a locally closed subset of $((\mathbb{P}^2)^\vee)^n$, resp. of the Grassmannian $\mathrm{G}(3, n)$, called the *configuration space* $\mathrm{Conf}(P)$. This is not the moduli space of stable pairs that we are interested in, but it is certainly related to it.

A theorem of Mnev says that, for any affine scheme of finite type Y over $\mathbb{Z}$, there exist n and P such that Y is locally analytically isomorphic to $\mathrm{Conf}(P)$ modulo smooth factors.

In other words, any singularity over $\mathbb{Z}$ or over the base field k that can be written using finitely many polynomials with integral coefficients in finitely many variables appears on one of the configuration spaces. This was used in [57] to show that many familiar moduli spaces have arbitrarily bad singularities. This includes the moduli spaces of *smooth* surfaces of general type. Basically, one needs to construct a class of varieties with are in bijection with hyperplane arrangements of type P, for example by taking Galois covers over a line arrangement in $\mathbb{P}^2$.

This principle of arbitrarily bad singularities most likely applies to the moduli spaces of weighted stable hyperplane arrangements of dimension bigger than or equal to two as well.

Chapter 5

Weighted Stable Hyperplane Arrangements

5.1 GIT and VGIT

We give a brief introduction to *Geometric Invariant Theory* (GIT, for short) and to variations of GIT quotients (VGIT, for short). In general, GIT is a big and nontrivial theory, for arbitrary reductive groups G. However, the main point of this introduction is that, when G is a torus, the GIT quotients are very simple, and computing them is an easy combinatorial exercise.

For a thorough introduction to GIT see [47]. For VGIT, see, e.g., [19].

5.1.1 Main definitions and results of GIT

In algebraic geometry, it is easy to take quotients of an algebraic variety by the action of a finite group G. One covers X by G-invariant open affine sets $U_i = \operatorname{Spec} R_i$. Then X/G is covered by the open affine sets $\operatorname{Spec} R_i^G$, where $R_i^G \subset R_i$ is the subring of invariants. The points of X/G are in bijection with the G-orbits of $G \curvearrowright X$.

This construction runs into immediate problems when G is infinite. The easiest example is $\mathbb{G}_m \curvearrowright \mathbb{A}^1$. If the orbits corresponded to points, then $\mathbb{A}^1/\mathbb{G}_m$ would have two points, and one would lie in the closure of the other. Moreover, it does not help to work with schemes here instead of varieties.

The basic definitions to handle the general case are the following.

Definition 5.1.1. The action of an algebraic group G on a variety X is the morphism $a: G \times X \to X$ satisfying the axioms of the group action.

Definition 5.1.2. A *categorical quotient* is a variety Y with a trivial G-action and with a G-equivariant (i.e., commuting with the G-action) morphism $f: X \to Y$ which has a universality property: for any other such pair $(Y', f': X \to Y')$, f' factors uniquely through f.

As any other object defined by a universality property, the pair (Y, f) is unique up to a canonical isomorphism.

Definition 5.1.3. A *geometric quotient* is a variety Y with a trivial G-action and with a G-equivariant morphism $f: X \to Y$ such that:

1. the k-points of Y are precisely the G-orbits on X;
2. a subset $U \subset Y$ is open if and only if $f^{-1}(U)$ is open; and
3. the regular functions on Y are precisely the G-invariant functions on X, i.e., for any open subset $U \subset Y$, $\Gamma(U, \mathcal{O}_Y) = \Gamma(f^{-1}(U), \mathcal{O}_X)$.

In the example with $\mathbb{G}_m \curvearrowright \mathbb{A}^1$ the geometric quotient does not exist, but the categorical quotient is a point $\operatorname{Spec} k$. When a geometric quotient exists, it is also a categorical quotient.

In GIT, one always works with an infinite *reductive* group, such as a multiplicative torus or SL_n, GL_n, SP_n. The first result of GIT is for the affine case:

Theorem 5.1.4.   1. *If $X = \operatorname{Spec} R$ then the categorical quotient exists and it is equal to $Y = \operatorname{Spec} R^G$, where $R^G \subset R$ is the ring of invariants.*

2. *The points of Y are identified with G-orbits on X modulo the following equivalence relation: $G \cdot x_1 \sim G \cdot x_2 \iff \overline{G \cdot x_1} \cap \overline{G \cdot x_2} \neq \varnothing$.*

3. *Among equivalent orbits, there exists a unique closed orbit which is contained in the closure of any other orbit in this equivalence class.*

Example 5.1.5. Consider $\mathbb{A}^2$ with two different actions by the group $G = \mathbb{G}_m$ illustrated in Figures 5.1 and 5.2. (If it helps, the reader may work over $\mathbb{R}$ and think of G as $\mathbb{R}^*$.)

1. Consider the action $\lambda.(x, y) = (\lambda x, \lambda^{-1} y)$. In terms of characters of G, we have $\operatorname{wt}(x) = 1$, $\operatorname{wt}(y) = -1$.

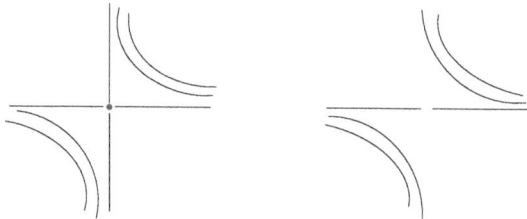

Figure 5.1: $\mathbb{G}_m \curvearrowright \mathbb{A}^2$ with weights $1, -1$.

The ring of invariants is $k[xy]$, so $\mathbb{A}^2/G = \mathbb{A}^1$. For $c \neq 0$, the orbit $xy = c$ is closed and gives a point of $\mathbb{A}^2/G$. For $c = 0$, the set $xy = 0$ consists of orbits $x = 0, y \neq 0$, $y = 0, x \neq 0$, and $x = y = 0$. The last orbit $x = y = 0$ is closed, and the other ones are equivalent to it. So, the three orbits get identified in the quotient.

Now remove the line $x = 0$. The new variety is $\mathbb{A}^2 \setminus \mathbb{A}^1 = \operatorname{Spec} k[x, 1/x, y]$. The ring of invariants is still $k[xy]$, so $(\mathbb{A}^2 \setminus \mathbb{A}^1)/G = \mathbb{A}^1$ is the same as before. This time, all orbits are closed.

2. Consider the action $\lambda.(x,y) = (\lambda x, \lambda y)$. In terms of characters of G, we have $\mathrm{wt}(x) = 1$, $\mathrm{wt}(y) = 1$.

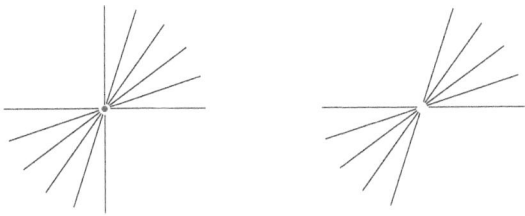

Figure 5.2: $\mathbb{G}_m \curvearrowright \mathbb{A}^2$ with weights $1, 1$.

The ring of invariants is k, so $\mathbb{A}^2/G = pt$. The orbits are $y = cx$ and $x = 0$. After removing the line $x = 0$, the ring of invariants is $k[x, 1/x, y]^G = k[y/x]$, so $(\mathbb{A}^2 \setminus \mathbb{A}^1)/G = \mathbb{A}^1$, and all the orbits are closed. This shows that removing some orbits may result into a bigger quotient!

The second main result is for the case when X is a projective variety. The main idea is very simple: if (X, L) is a polarized projective variety and L is a G-linearized ample line bundle, then G acts on the ring $R(X, L) = \bigoplus_{d \geq 0} H^0(X, L^d)$. The group acts on the affine cone $\widetilde{X} = \mathrm{Spec}\, R(X, L)$ and the categorical quotient $\widetilde{X}/G$ is $\widetilde{Y} = \mathrm{Spec}\, R(X, L)^G$. The variety X is the projective version $X = \mathrm{Proj}\, R(X, L)$, so the quotient of X should be $\mathrm{Proj}\, R(X, L)^G$.

However, there is one point of the affine cone $\widetilde{Y}$ which does not give a point in Y, namely the vertex 0. Therefore, one must remove the G-orbits in $\widetilde{X}$ equivalent to 0, i.e., the orbits in $\widetilde{Y}$ with $0 \in \overline{G \cdot \widetilde{x}}$.

Definition 5.1.6. A point $x \in X$ is called *unstable* if for the corresponding points $\widetilde{x} \in \widetilde{X}$ one has $0 \in \overline{G \cdot \widetilde{x}}$. Let X^{uns} be the set of all unstable points.

A point $x \in X$ is called *semistable* if $x \in X \setminus X^{\mathrm{uns}}$. The set of all semistable points is denoted X^{ss}.

Finally, a point $x \in X$ is called *stable* if $x \in X^{\mathrm{ss}}$, the orbit $G \cdot x \subset X^{\mathrm{ss}}$ is closed and the stabilizer G_x is finite. The set of all stable points is denoted X^{s}.

The main result of GIT is the following:

Theorem 5.1.7. *Let (X, L) be a polarized projective variety with a G-linearized ample line bundle L. Then*

1. *the set X^{ss} is open in X, and $\mathrm{Proj}\, R(X, L)^G$ is its categorical quotient;*
2. *the points of X^{ss}/G are G-orbits of X^{ss} modulo the equivalence relation $G \cdot x_1 \sim G \cdot x_2 \iff \overline{G \cdot x_1} \cap \overline{G \cdot x_2} \neq \varnothing$;*
3. *among the equivalent orbits, there exists a unique closed orbit which is contained in the closure of any other orbit in this equivalence class;*
4. *the set X^{s} is open in X^{ss} and its geometric quotient exists; the points of X^{s}/G are G-orbits of X^{s}.*

The categorical quotient of X^{ss} is denoted by $X/\!\!/G$. It bears repeating that X^{ss}, X^s, and $X/\!\!/G$ depend on the choice of a G-linearized ample line bundle L.

Example 5.1.8. Consider the same actions as in Example 5.1.5, but this time consider the corresponding projective variety $(\mathbb{A}^2) = \mathbb{P}^1$ with a linearized ample line bundle $\mathcal{O}(1)$.

1. The unstable locus is $xy = 0$, $X^{ss} = X^s = \mathbb{P}^1 \setminus \{(0,1),(1,0)\}$, and $X/\!\!/G$ is a point.

2. The semistable locus is empty and $X/\!\!/G = \varnothing$.

The most general statement of GIT [47, Thm.1.1.10] is for arbitrary Noetherian schemes X with an arbitrary G-linearized sheaf. Again, one defines the open sets $X^{ss} \supset X^s$, but with trickier definitions, and the main results assert the existence of the categorical quotient $X^{ss}/\!\!/_G$ and of the geometric quotient X^s/G.

5.1.2   GIT quotient by a torus action

Now suppose that G is a torus $T = \mathbb{G}_m^r$. The action $T \curvearrowright R = R(X,L)$ is diagonalizable and decomposes R into a direct sum $R = \bigoplus_{m \in M} R_m$. Then the ring of invariants R^G is simply R_0, the degree-0 part. Thus, from the algebraic point of view, the GIT quotient by a torus action is extremely easy: $X/\!\!/T = \operatorname{Proj} R_0$.

This becomes especially easy for toric varieties. Let (X, L) be a polarized toric variety for a big torus $H = \mathbb{G}_m^N$ with an H-linearized ample line bundle L. Recall that it corresponds to a polytope Q. Explicitly, $X = \operatorname{Proj} R$, where $R = k[\mathbb{Z}^N \cap \operatorname{Cone}(1, Q)]$ and $L = \mathcal{O}(1)$ on this Proj. The ring R is graded by $\mathbb{Z} \oplus M_H$ and, in particular, by M_H.

Let $T \subset H$, $T = \mathbb{G}_m^r$ be a subtorus. On the character lattices we have a surjection $\phi \colon M_H \to M_T$. This gives a grading on R by the lattice M_T. Now the ring of invariants R^T, i.e., the 0-degree part of R is simply

$$k[\mathbb{Z}^N \cap \operatorname{Cone}(1, Q')], \qquad Q_0 = Q \cap \phi^{-1}(0).$$

Thus, it corresponds to a slice $Q \cap \phi^{-1}(0)$ of the polytope Q. This is illustrated in Figure 5.3.

5.1.3   Variation of GIT quotients (VGIT)

VGIT is best illustrated by the above situation, as in Figure 5.3. First of all, replacing the line bundle L by a multiple L^a results in replacing the polytope Q by a multiple aQ. The ring $R = \bigoplus_{d \geq 0} H^0(X, L^d)$ is replaced by the Veronese subring $R^{(a)} = \bigoplus_{d \geq 0} H^0(X, L^{ad})$. One has $\operatorname{Proj} R = \operatorname{Proj} R^{(a)}$, so the variety X is unchanged. Thus, for the GIT quotient purposes, one can freely replace L by a positive multiple.

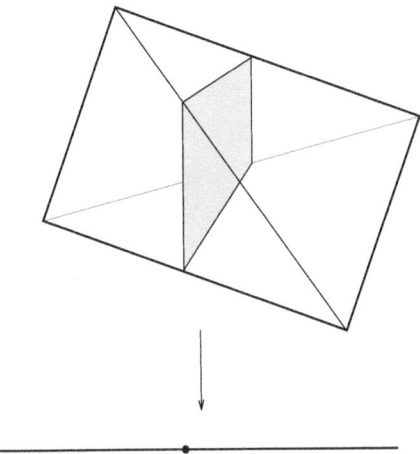

Figure 5.3: GIT quotient of a polarized toric variety.

This allows us to talk about fractional T-linearizations of L. By definition, it is a T-linearization of some multiple L^a. Now changing the original linearization in combinatorial terms amounts to replacing the slice $Q_0 = Q \cap \phi^{-1}(0)$ by a parallel slice $Q_c = Q \cap \phi^{-1}(c)$, $c \in M_T \otimes \mathbb{Q}$.

Now pick a generic c in Figure 5.3. Then for a nearby c' the polytope $Q_{c'}$ is *normally equivalent* to Q_c. This means that the combinatorics of the faces of Q_c and $Q_{c'}$ are the same, and one is obtained from the other by parallel shift of the facets. More precisely, this means that the normal fans of Q_c and $Q_{c'}$ are the same. So, the associated toric varieties $Y_c = X /\!/_c T$ and $Y_{c'} = X /\!/_{c'} T$ are also the same.

However, if one considers some special c, then the combinatorics change and $Y_{c'}$ no longer equals Y_c. Nevertheless, there is still a contraction $Y_{c'} \to Y_c$. When Q_c is maximal-dimensional, it is a birational contraction. When c is a boundary point then it is a projective morphism with positive-dimensional fibers.

Finally, when c lies outside of the projection $\phi(Q)$, we have $Q_c = \varnothing$, and so the quotient Y_c is also empty. Putting this together, we get the following result.

Theorem 5.1.9. 1. *The set of $\mathbb{Q}$-linearizations c of L is divided into finitely many polyhedral chambers.*

2. *If c, c' lie in the same locally closed chamber, then $Y_c = Y_{c'}$.*

3. *If c is a specialization of c', denoted $c' \in \bar{c}$, then there exists a proper contraction $\pi \colon Y_{c'} \to Y_c$ with $\pi_* \mathcal{O}_{Y_{c'}} = \mathcal{O}_{Y_c}$.*

This theorem is one-half of VGIT. The second half consists in varying L in $\operatorname{Pic} X \otimes \mathbb{Q}$, and is equally easy.

The only strengthening of this simple VGIT that is needed for weighted hyperplane arrangements is the following one. Let (X, L) be a polarized variety with a G-linearized ample line bundle L. Let $Z \subset X$ be a closed G-invariant subvariety. Then the GIT quotient $Z /\!/ G$ w.r.t. the G-linearized ample line bundle $L|_Z$ is a closed subvariety of $X /\!/ G$.

This simple observation allows one to extend the above VGIT picture from toric varieties to a much larger class of T-invariant subvarieties of toric varieties.

5.2   Semi-log canonical singularities and GIT

The following theorem is the heart of the construction of weighted stable hyperplane arrangements. This is where the magic happens.

Recall that a point of a Grassmannian $G(r, n) \subset \mathbb{P}^N$, $N = \binom{n}{r} - 1$, is a linear space $\mathbb{P}V \subset \mathbb{P}^{n-1}$ and, assuming $\mathbb{P}V$ does not lie in any of the coordinate hyperplanes H_i, $i = 1, \ldots, n$, we have a hyperplane arrangement $(\mathbb{P}V, B_i = H_i \cap \mathbb{P}V)$.

Thus, we have a universal family $P \to G(r, n)$, $P \subset \mathbb{P}^N \times \mathbb{P}^{n-1}$ whose fibers are the linear spaces $\mathbb{P}V$. Note that $\mathbb{P}^N \times \mathbb{P}^{n-1}$ is a toric variety for the torus $H = \mathbb{G}_m^{N+n-1}$, that $T \subset H$ is a subtorus, and that $G(r, n)$ and P are T-invariant subvarieties of $\mathbb{P}^N$ and $\mathbb{P}^N \times \mathbb{P}^{n-1}$, respectively.

Now pick a vector $\boldsymbol{b} = (b_1, \ldots, b_n)$, $0 < b_i \le 1$, $b_i \in \mathbb{Q}$. Assume that $\boldsymbol{b}(\overline{n}) > r$. To this vector we associate both an ample $\mathbb{Q}$-line bundle $L_{\boldsymbol{b}} = \mathcal{O}(1, \boldsymbol{b}(\overline{n}) - r)$, and a T-linearization of $L_{\boldsymbol{b}}$. This last one is defined by setting the weight of the variable z_i to be

$$\mathrm{wt}(z_i) = \boldsymbol{e}_i - \frac{\boldsymbol{b}}{\boldsymbol{b}(\overline{n})} \in M_T \otimes \mathbb{Q},$$

for $i = 1, \ldots, n$. This gives a choice of a linearized ample line bundle on $\mathbb{P}^N \times \mathbb{P}^{n-1}$ and its subvariety P. To emphasize this choice, we will talk about $\boldsymbol{b}$-semistable points, the quotient $P /\!/_{\boldsymbol{b}} T$, etc.

Theorem 5.2.1 ([7]). *Let $[V] \in G(r, n)$ be a point in the Grassmannian and let $[p \in V] \in P$ be a point in the universal family over it. The following holds:*

1. *if $V \subset \bigcup_{i=1}^n H_i$ then $[p \in V]$ is not $\boldsymbol{b}$-semistable;*

2. *if $V \not\subset \bigcup_{i=1}^n H_i$ then $[p \in V]$ is $\boldsymbol{b}$-semistable if and only if $\mathrm{BP}_V \cap \Delta_{\boldsymbol{b}} \ne \varnothing$ and the pair $(\mathbb{P}V, \sum b_i B_i)$ is log canonical at p.*

There is a general principle that log canonical and semi-log canonical properties are very close to GIT stability, and generally GIT semistable is stronger. See [26] for a discussion and [49] for a recent example. Theorem 5.2.1 is an instance of this principle, and it works especially nicely because hyperplanes B_i are linear subvarieties.

5.3   Weighted shas

Consider a matroid tiling $\bigcup \mathrm{BP}_{V_i}$ of Δ_b. Recall from Definition 4.5.2 that this means the following:

1. Each polytope BP_{V_i} is the base polytope of some hyperplane arrangement. Polytopes of maximal dimension $n-1$ correspond to arrangements with trivial automorphism group, polytopes of codimension c correspond to arrangements with $\mathrm{Aut}(\mathbb{P}^{r-1}, B_1, \ldots, B_n) = (k^*)^c$.

2. The tiling $\bigcup \mathrm{BP}_{V_i}$ is face-fitting: any two polytopes are either disjoint, or intersect along a smaller base polytope, with bigger automorphism group, which is a face of both.

3. Each BP_{V_i} intersects the b-cut hypersimplex $\Delta_b(r, n)$.

4. $\bigcup \mathrm{BP}_{V_i} \supset \Delta_b(r, n)$, but they do not have to cover the entire $\Delta(r, n)$.

5. Finally, recall that we ignore the polytopes lying entirely in the hyperplanes $x_s = 0$ because they do not correspond to hyperplane arrangements, since their matroids have loops (zero vectors). Thus, a more accurate but cumbersome statement would be that $\bigcup(\mathrm{BP}_{V_i} \smallsetminus \bigcup_{s=1}^n \{x_s = 0\}) \supset \Delta_b \smallsetminus \bigcup_{i=1}^n \{x_i = 0\}$.

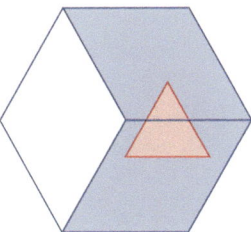

Figure 5.4: A matroid tiling of the b-cut hypersimplex Δ_b.

Now, assume that $Y \to \mathrm{G}(r, n)$ is a stable toric variety of type $\bigcup \mathrm{BP}_{V_i}$. Thus:

1. $Y = \cup Y_i$, and each irreducible component $Y_i = \overline{T.V_i}$ is the closure of an orbit of an arrangement $\mathbb{P}V_i \subset \mathbb{P}^{n-1}$ in $\mathrm{G}(r, n)$, with base polytope BP_{V_i}.

2. They are glued along smaller orbits: $\overline{T.V_i} \supset \overline{T.V_j}$ means that $[V_j] = \lim_\lambda [V_i]$ for some 1-parameter subgroup $\lambda \colon \mathbb{G}_m \to T$.

3. By 4.1.6, each Y_i is a closed subvariety of $\mathrm{G}(r, n)$ so, $Y \to \mathrm{G}(r, n)$ is a closed embedding.

Let $P \to \mathrm{G}(r, n)$ be the universal family with fibers $\mathbb{P}V \simeq \mathbb{P}^{r-1}$ and let $P_Y = P \times_{\mathrm{G}(r,n)} Y$ be its restriction to Y. We have $P_Y \subset \mathbb{P}^{n-1} \times Y$.

Definition 5.3.1. A *weighted stable hyperplane arrangement* (*sha*, for short) associated to the stable toric variety Y over $\mathrm{G}(r, n)$ for the weight $b = (b_1, \ldots, b_n)$ is the GIT quotient $X = P_Y /\!\!/ T$, together with the hyperplanes $B_i = (P_Y \cap H_i) /\!\!/_b T$.

Theorem 5.3.2 (Main theorem). *The pair* $(X, \sum b_i B_i)$ *has semi-log canonical singularities, and the* $\mathbb{Q}$-*divisor* $K_X + \sum b_i B_i$ *is ample. In other words,* $(X, \sum b_i B_i)$ *is a stable pair.*

The proof of this theorem is a combination of the following two ingredients: by Lemma 2.6.7, the stable toric variety Y has slc singularities, and, by Theorem 5.2.1, for every semistable point $[p \in \mathbb{P}V]$ in the fiber $P_Y \to Y$, the fiber $(\mathbb{P}V, \sum b_i B_i)$ is lc.

Example 5.3.3. Suppose that there is a single polytope BP_V covering Δ_b, so that $\Delta_b \subset \mathrm{BP}_V^0$. Then, by Theorem 5.2.1, the semistable points in P_Y are of the form $[p \in \mathbb{P}(t.V)]$, where $t.V \in \mathrm{G}(r, n)$ is any of the linear spaces in the T-orbit of $V \subset \mathbb{P}^{n-1}$ and p is an arbitrary point of $\mathbb{P}(t.V)$. Since BP_V is maximal-dimensional, the stabilizer of V is free and $T.V \simeq V$.

The action of the torus on the orbit $T.[V]$ in $\mathrm{G}(r, n)$ is free so, moreover, the orbits of the points $[p \in \mathbb{P}V]$ are free. Hence, the quotient is $[\mathbb{P}V, B_1, \ldots, B_n]$, and $(X, \sum b_i B_i) = (\mathbb{P}V, \sum b_i B_i)$.

Example 5.3.4. Consider the subdivision of $\Delta(3, 5)$ into $\{x_{123} \le 2\}$ and $\{x_{45} \le 1\}$, and their faces. Let $\boldsymbol{b} = (1 - \epsilon, \ldots, 1 - \epsilon)$, so we don't have to worry about the faces $x_i = 1$ for now. Then the tiling of Δ_b consists of three polytopes: $\mathrm{BP}_{V_1} = \{x_{123} \le 2\}$, $\mathrm{BP}_{V_2} = \{x_{45} \le 2\}$, and $\mathrm{BP}_{V_3} = \{x_{123} = 2, x_{45} = 1\}$.

The set of the semistable point consists of the T-orbits of the following points: $\mathbb{P}V_1 \setminus p_{123} = \mathbb{P}^2$ minus a point, $\mathbb{P}V_2 \setminus \ell_{45} = \mathbb{P}^2$ minus a line, and $\mathbb{P}V_3 \setminus (p_{123} \cup \ell_{45}) = \mathbb{P}^2$ minus a point and a line.

The actions on the orbits $T \cdot [V_1]$ and $T \cdot [V_2]$ are free. The action on $T \cdot [V_3]$ is *not* free, the stabilizer is $\mathbb{G}_m$. However, the action on the orbits of the points $[p \in \mathbb{P}V_3]$ is free, since $\mathrm{Aut}(\mathbb{P}^2, B_1, \ldots, B_5, p) = 1$. Therefore, the quotient of the above set is the union of $\mathbb{P}V_1 \setminus p_{123}$, $\mathbb{P}V_2 \setminus \ell_{45}$ and $(\mathbb{P}V_3 \setminus \{p_{123} \cup \ell_{45}\})/\mathbb{G}_m = \mathbb{P}^1$. It is easy to see that X is the union of $\mathrm{Bl}_1 \mathbb{P}^2 = \mathbb{F}_1$ and $\mathbb{P}^2$ joined along a line $\mathbb{P}^1$.

Theorem 5.3.5. *For any* $\boldsymbol{b}$ *and any* $0 < \epsilon \ll 1$, *let* $\boldsymbol{b}' = \boldsymbol{b} - \epsilon$. *Then,*

1. $Y_{\boldsymbol{b}'}^{\mathrm{ss}} = Y_{\boldsymbol{b}'}^{\mathrm{s}}$, *and the action on* $Y_{\boldsymbol{b}'}^{\mathrm{s}}$ *is free; moreover, the variety* $X_{\boldsymbol{b}'} = Y_{\boldsymbol{b}'}^{\mathrm{s}}/T$ *is a geometric quotient;*

2. *there exists a contraction* $\pi \colon X_{\boldsymbol{b}'} \to X_{\boldsymbol{b}}$ *which is crepant w.r.t* $\boldsymbol{b}$, *i.e.,*
$$K_{X_{\boldsymbol{b}'}} + \sum b_i B_i' = \pi^*(K_{X_{\boldsymbol{b}}} + \sum b_i B_i);$$

3. *if, additionally,* N *is a positive integer such that all* $Nb_i \in \mathbb{Z}$, *then for any weighted sha w.r.t.* $\boldsymbol{b}$ *the divisor* $N(K_X + \sum b_i B_i)$ *is Cartier;*

4. *the morphism* $\pi \colon X_{\boldsymbol{b}'} \to X_{\boldsymbol{b}}$ *is birational (on every irreducible component), and it is an isomorphism over* $\bigcup B_i$.

The facts that $Y_{\boldsymbol{b}'}^{\mathrm{ss}} = Y_{\boldsymbol{b}'}^{\mathrm{s}}$ for a general $\boldsymbol{b}'$, and the existence of a contraction of GIT quotients $\pi \colon X_{\boldsymbol{b}'} \to X_{\boldsymbol{b}}$ are standard properties of VGIT.

The singularities of a quotient $Y_{\boldsymbol{b}'}^{\mathrm{s}}/T$ are the same as the singularities of $Y_{\boldsymbol{b}'}^{\mathrm{s}}$. This, together with applying the contraction π from the above theorem, implies the following results.

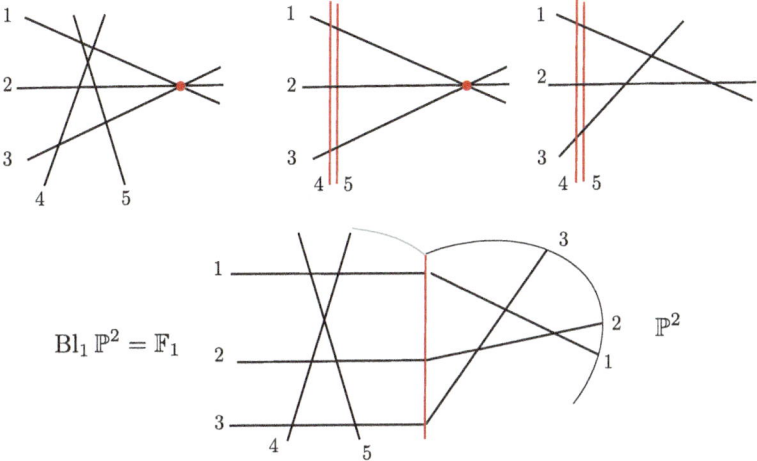

Figure 5.5: Illustration for Example 5.3.4.

Theorem 5.3.6. *The following holds:*

1. *for a generic $\boldsymbol{b}'$, the variety $X_{\boldsymbol{b}'}$ is Gorenstein;*
2. *for any $\boldsymbol{b}$, the variety $X_{\boldsymbol{b}}$ is Cohen–Macaulay, and $X_{\boldsymbol{b}} \smallsetminus \bigcup B_i$ is Gorenstein.*

Theorem 5.3.7. *The finer structure of $X = \bigcup X_s$ is described by the following facts:*

1. *the stratification of X into irreducible components, and their intersections (we do not include the divisors B_i into this) coincides with the stratification of the polytopal complex $\Delta_{\boldsymbol{b}} = \bigcup (\mathrm{BP}_{V_s} \cap \Delta_{\boldsymbol{b}})$;*
2. *every irreducible component of X_s of $X_{\boldsymbol{b}}$ is normal; in fact, it is the log canonical model of the non-lc hyperplane arrangement $(\mathbb{P}V_s, \sum b_i B_{i,s})$;*
3. *for every irreducible component X_s, the open subset $X_s \smallsetminus (\bigcup_{j \neq s} X_j \cup B_i)$ is isomorphic to $\mathbb{P}V_s \smallsetminus \bigcup B_i$ for the corresponding hyperplane arrangement.*

5.4 Moduli spaces of shas

We start with the open part of the moduli space we intend to compactify.

Definition 5.4.1. Fix positive integers r, n and let $\boldsymbol{b} = (b_1, \ldots, b_n)$ be a vector with $0 < \boldsymbol{b} \leq 1$ and $\boldsymbol{b}(\bar{n}) > r$. Let $\mathrm{M}_{\boldsymbol{b}}$ be the moduli space of log canonical hyperplane arrangements $(\mathbb{P}^{r-1}, \sum_{i=1}^{n} b_i B_i)$.

The space $\mathrm{M}_{\boldsymbol{b}}$ is fairly easy to construct, as follows. Let $U \subset \mathrm{G}(r, n)$ parameterize the pairs such that $(X, \sum b_i B_i)$ is log canonical. Then U is an open subset in a smooth variety contained in the set $\mathrm{G}(r, n)_{\boldsymbol{b}}^s$ of stable points for the linearization defined by $\boldsymbol{b}$. Thus, there exists a geometric quotient $\mathrm{M}_{\boldsymbol{b}} = U/G \subset \mathrm{G}(r, n)/\!/_{\boldsymbol{b}} T$.

Note that a log canonical pair has trivial automorphisms, for example, because by Theorem 4.4.2 the base polytopes BP_V are maximal-dimensional. Thus, M_b is smooth, and it is a fine moduli space. It is easy to compute its dimension as

$$\dim M_b = n(r-1) - (r^2-1) = (r-1)(n-r-1).$$

For example, $\dim M_b(2,n) = n-3$ and $\dim M_b(3,n) = 2(n-4)$.

Theorem 5.4.2. *For every r, n and $b = (b_1, \ldots, b_n)$, there exist a projective scheme $\overline{M}_b(r,n)$ and a flat projective family $(\mathcal{X}, \mathcal{B}_1, \ldots, \mathcal{B}_n) \to \overline{M}_b(r,n)$ such that every fiber is one of the weighted shas defined in Section 5.3 (thus, $(X, \sum b_i B_i)$ is slc and $K_X + \sum b_i B_i$ is ample), and there are no repeating fibers.*

An individual weighted sha was constructed starting from a stable toric variety over the Grassmannian $G(r,n)$. Thus, to construct a family of shas one has to consider a family of stable toric varieties, as in Theorem 2.4.3. This works well for the weight $b = 1$ considered in [27]. For an arbitrary weight b, however, we have a problem: the cover of Δ_b was only a *partial* cover of $\Delta(r,n)$, and partial covers for different supports can not vary flatly: the Hilbert polynomial changes.

Combinatorially, it is clear what is going on: the partial cover $\cup \mathrm{BP}_{V_i}$ of $\Delta(r,n)$ is irrelevant, the only important part is the cover of $\Delta_b(r,n)$ by the polytopes $\mathrm{BP}_{V_i} \cap \Delta(r,n)$. And for these covers the topological type, the support Δ_b is the same, and one can freely apply Theorem 2.4.3 to an appropriate multiple $N\Delta_b$ to make sure that it is a lattice polytope.

The solution is to replace the Grassmannian by a b-cut Grassmannian $G_b(r,n)$ whose moment polytope is Δ_b. We will not go into details of this construction here, referring an interested reader to [7].

5.5 Geography of the moduli spaces of shas

In the title of the present section, "geography" refers to varying the weight b. How are the moduli spaces $\overline{M}_b$ related for different b? What is located in the extreme corners? The answer is the following theorem, illustrated in Figure 5.6.

Definition 5.5.1. The weight domain of possible weights b is

$$\mathcal{D}(r,n) = \left\{ b \in \mathbb{Q}^n \mid 0 < b_i \leq 1, \ \sum b_i > r \right\}.$$

The closure $\overline{\mathcal{D}(r,n)}$ is a polytope whose lower face is

$$\{ x \in \mathbb{R}^n \mid 0 \leq x \leq 1, \ x(\overline{n}) = r \} = \Delta(r,n).$$

Theorem 5.5.2. *The domain $\mathcal{D}(r,n)$ is divided by the hyperplanes $x(I) = k$ for all $I \subset \overline{n}$, $2 \leq |I| \leq n-2$, $1 \leq k \leq n-1$, into finitely many chambers. Then,*

1. *(same chamber) if b and b' lie in the same chamber, denoted $b \sim b'$, then $\overline{M}_b = \overline{M}_{b'}$ and the families of shas are the same;*

2. (*specialization*) *if* $b \in \overline{\mathrm{Chamber}(b')}$, *denoted* $b \in \overline{b'}$, *then there exists a contraction* $\overline{M}_{b'} \to \overline{M}_b$ *on the moduli spaces and* $(\mathcal{X}', \mathcal{B}'_i) \to (\mathcal{X}, \mathcal{B}_i)$ *on the families;*

3. (*specialization from below*) *further, if* $b \in \overline{b'}$ *and* $b' \le b$, *then* $\overline{M}_{b'} = \overline{M}_b$ *and on the fibers the morphism* $X' \to X$ *is birational (on every irreducible component).*

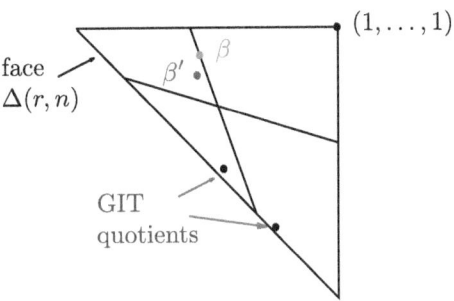

Figure 5.6: Chamber decomposition for the weights b.

Theorem 5.5.3. *Let* $a \in \Delta(r,n)$ *be a generic element of the lower face and* $b \in \mathcal{D}(r,n)$ *be an element such that* $a \in \overline{b}$. *Then* $\overline{M}_b = M_b^0 = \mathrm{G}(r,n)/\!\!/_a T$, *i.e., all weighted shas in this case are ordinary log canonical hyperplane arrangements* $(\mathbb{P}^{r-1}, \sum b_i B_i)$.

5.6 Shas of dimension 1

Let $(X, \sum b_i B_i)$ be a weighted sha of dimension 1. By Theorem 5.3.7, every irreducible component X_s is normal and birationally isomorphic to $\mathbb{P}^1$, so it is a $\mathbb{P}^1$. The stratification of X into irreducible component coincides with the stratification of Δ_b into $\cup (\mathrm{BP}_{V_s} \cap \Delta_b)$. Now Theorem 4.6.2 about partial tilings of $\Delta(2,n)$ gives the following result.

Theorem 5.6.1. *A weighted sha of dimension 1 is a tree of* $\mathbb{P}^1$ *'s. The faces* $x(K_a) \le 1$ *correspond to points* $Q_a \in X$ *distinct from the nodes and the points* $B_i = Q_a$ *for* $i \in K_a$.

Thus, the one-dimensional shas are the same as the weighted stable curves introduced in Hassett [30]. The chamber decomposition from Theorem 5.5.2 is also the same as Hassett's chamber decomposition. The hyperplanes are of the form $x(I) = 1$ for $I \subset \overline{n}$.

Finally, a birational morphism of one-dimensional varieties is an isomorphism, so this provides an additional simplification in the one-dimensional case: for $b \in \overline{b'}$, $b' \le b$ (specialization from below) not only $\overline{M}_{b'} = \overline{M}_b$ but also $X' = X$.

The geography in this case is very easy to describe. When changing the weights, every time we cross a wall $\boldsymbol{b}(I) = 1$, $|I| \geq 2$, downwards, $\boldsymbol{b}' > \boldsymbol{b}$, $\boldsymbol{b}(I) > 1$ to $\boldsymbol{b}'(I) \leq 1$, the pair $(X, \sum b_i B_i)$ shown in the Figure 5.7 ceases to be stable because the degree of $K_X + \sum b_i B_i$ on the end component E, which by adjunction equals

$$\deg K_{\mathbb{P}^1} + \#(\text{double points on } E) + \sum_{B_i \in E} b_i = -1 + \sum_{i \in I} b_i,$$

goes from positive to non-positive. This end component on $X_{b'}$ gets contracted to yield the new curve X_b on which the points B_i coincide for $i \in I$.

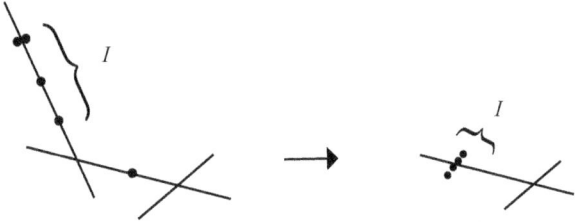

Figure 5.7: Crossing a wall $\boldsymbol{b} = 1$.

5.7   Shas of dimension 2

5.7.1   General results

Theorem 5.3.7 together with the fact that $(K_X + \sum b_i B_i)|_{X_s}$ is ample gives enough control over irreducible components of X.

Definition 5.7.1. Recall that lines B_i in a hyperplane arrangement are allowed to coincide. Denote by $\operatorname{Simp}(\mathbb{P}V, \sum B_i)$ the hyperplane arrangement where the coinciding lines $B_I = B_i$, $i \in I$, are counted once with the weight $b'_I = \min(\sum_{i \in I} b_i, 1)$.

By applying Theorem 5.3.7, one obtains the following result.

Theorem 5.7.2. *For any weight $\boldsymbol{b}$, an irreducible component of a weighted sha is*

1. *either a blowup of $\mathbb{P}^2$ at $k \geq 0$ points, where the hyperplane arrangement $\operatorname{Simp}(\mathbb{P}V, \sum b_i B_i)$ is not lc,*

2. *or $\mathbb{P}^1 \times \mathbb{P}^1$ which is obtained from $\operatorname{Bl}_2 \mathbb{P}^2$ by contracting the (-1)-curve. This last case appears only for the hyperplane arrangement given in Figure 5.8 and only if $\operatorname{Simp}(X, \sum b_i B_i)$ is not lc at exactly two points p_1, p_2, and the line between them has weight 1 in $\operatorname{Simp}(X, \sum b_i B_i)$.*

The irreducible components are glued together along the following $\mathbb{P}^1$'s: the strict preimages in $\mathrm{Bl}\,\mathbb{P}^2$ of one-dimensional non-lc line loci of $(\mathbb{P}V, \sum b_i B_i)$, and the exceptional divisors of blowups, with the exception of the line in the case of Figure 5.8 which gets contracted.

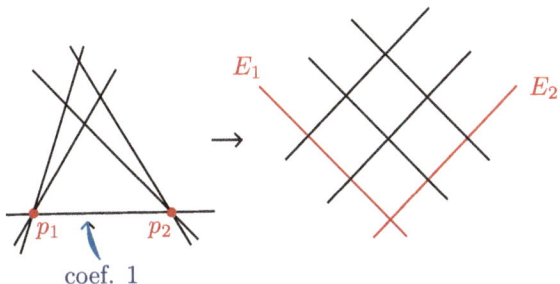

Figure 5.8: Line arrangement leading to $\mathbb{P}^1 \times \mathbb{P}^1$.

Proof. Indeed, the procedure for constructing the log canonical model for a non-lc pair $(\mathbb{P}V_s, \sum b_i B_{i,s})$ was the following:

1. Write $\sum b_i B_{i,s}$ as a the sum $\sum d_k D_k$ with distinct hyperplanes D_k, so that $d_k = \sum_{i,\ B_{i,s}=D_k}$. For the hyperplanes with weight $d_k > 1$, set $d'_k = 1$, for the others leave $d'_k = d_k$.

2. Blow up the non-lc points of the new arrangement, and call the exceptional divisors (-1)-curves E_j.

3. Apply MMP to $K + \sum d'_k f_*^{-1} D_k + \sum E_j$. In dimension two this means that we contract several curves.

By Theorem 5.3.7, the curves contracted by the MMP belong to the set $\{f_*^{-1} D_k\}$, so we only have to pay attention to them.

It is easy to see that, after the second step, the divisor $\sum d'_k f_*^{-1} D_k + \sum E_j$ is already nonnegative on these curves, and the only curves where it can be zero are the (-1)-curves obtained by blowing up exactly two points on $\mathbb{P}^2$, as in the statement of the theorem. □

We can now explain the "volumes" of the polytopes used in Section 4.7. Obviously, every lattice polytope has the usual Euclidean volume, which can be normalized so that the smallest polytope has volume 1. But the "volume" we define below is much smaller and more convenient.

Let $(X, \sum B_i)$ be a sha for the weight $\boldsymbol{b} = \boldsymbol{1}$. The divisor $K_X + \sum B_i$ is ample, Cartier, and has the same numerical invariants as the corresponding divisor on $\mathbb{P}^2$. Therefore,

$$\left(K_X + \sum B_i\right)^2 = \sum\left((K_X + \sum B_i)|_{X_s}\right)^2 = (K_{\mathbb{P}^2} + nH)^2 = (n-3)^2.$$

By adjunction, $(K_X + \sum B_i)|_{X_s} = K_{X_s} + \sum B_i + D_s$, where D_s is the double locus, the intersection of X_s with the other irreducible components of X.

Definition 5.7.3. To each irreducible component we associate the positive integer $(K_{X_s} + \sum B_i + D_s)^2$ and call it the *volume* of the corresponding polytope.

Note that one can define a refined version of the volume, as $(K_X + \sum b_i B_i + D_s)^2$ for a sha with weight b if $\mathrm{BP}_{V_s} \cap \Delta_b \neq \varnothing$, and 0 otherwise. These refined volumes are polynomials of degree 2 and they add up to $(\sum b_i - 3)^2$. Naturally, these definitions can be given for any matroid polytope in $\Delta(r, n)$ that corresponds to a hyperplane arrangement.

Here is an application of the volume.

Lemma 5.7.4. *Any weighted sha of dimension r has at most $(n - r)^{r-1}$ irreducible components.*

The actual bound is smaller because it turns out to be impossible for all the pieces to have volume 1.

Next, we go through explicit examples for $n \leq 6$ divisors.

5.7.2   The case $n = 4$

Nothing here, move along. One has $\dim \overline{M}_b = 2(n - 4) = 0$ and there is a unique stable pair $(X, \sum b_i B_i)$ for as long as $\sum b_i > 3$, namely, $\mathbb{P}^2$ and four lines in general position.

5.7.3   The case $n = 5$

The two nontrivial tilings of Section 4.7.2 for $b = 1$ give us the varieties shown in Figures 5.9 and 5.10.

1. The tiling (3) $x_{123} \leq 2$; (1) $x_{45} \leq 1$. There are three cases:
   (a) $b_{123} > 2$ and $b_{45} > 1$; then $X = \mathrm{Bl}_1 \mathbb{P}^2 \cup \mathbb{P}^2$, shown in Figure 5.9.
   (b) $b_{123} \leq 2$ and $b_{45} > 1$; then the first component is contracted to a line, and $X = \mathbb{P}^2$.
   (c) $b_{123} > 2$ and $b_{45} \geq 1$; then the second component is contracted to a point, and $X = \mathbb{P}^2$.

   Note that $b_{123} + b_{45} = b(\overline{n}) > 3$, so one of these cases must hold.

2. The tiling (2) $x_{125} \leq 2$, $x_{345} \leq 2$; (1) $x_{34} \leq 1$; (1) $x_{12} \leq 1$.

   For $b = 1$, $X = X_1 \cup X_2 \cup X_3$, with $X_1 = \mathbb{P}^1 \times \mathbb{P}^1$, $X_2 = \mathbb{P}^2$, $X_3 = \mathbb{P}^2$, as shown in Figure 5.10 on the left.

   For $b = (1, 1, 1, 1, 1-\epsilon)$, $X = X_1 \cup X_2 \cup X_3$, with $X_1 = \mathrm{Bl}_3 \mathbb{P}^2$, $X_2 = \mathbb{P}^2$, $X_3 = \mathbb{P}^2$, as shown in Figure 5.10 on the right.

   When $b_{125} \leq 2$, X_2 gets contracted. When $b_{345} \leq 2$, X_3 gets contracted.

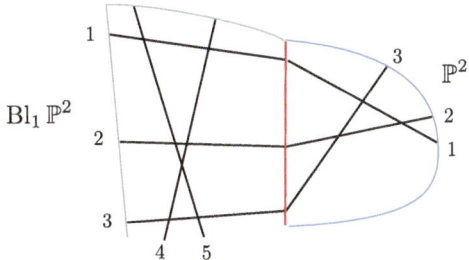

Figure 5.9: $n = 5$, tiling 1.

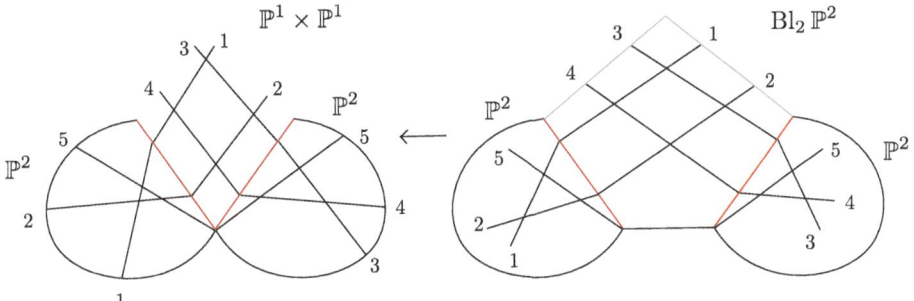

Figure 5.10: $n = 5$, tiling 2.

When $b_{12} \leq 1$, both X_1 and X_2 get contracted. When $b_{34} \leq 1$, both X_1 and X_3 get contracted.

Starting with the picture on the right, when either X_2 or X_3 or both are contracted, on $X_1 = \mathrm{Bl}_2 \, \mathbb{P}^2$ one or both of the exceptional curves are contracted.

Starting with the picture on the left with $b_5 = 1$, when either X_2 or X_3 is contracted, X_1 must also be contracted, because $b_5 = 1$, $b_{125} \leq 2$ implies $b_{25} \leq 1$. And both X_2 and X_3 can not get contracted at the same time because $b_5 = 1$, $b_{125} \leq 2$, $b_{345} \leq 2$ implies that $b(\overline{5}) \leq 3$.

Also, note that the specialization up $\boldsymbol{b}' = (b_1, \ldots, b_4, 1-\epsilon) \rightsquigarrow \boldsymbol{b} = (b_1, \ldots, b_4, 1)$, provided both $b_{125} > 2$ and $b_{345} > 2$, results in a nontrivial birational contraction $X_{\boldsymbol{b}'} \to X_{\boldsymbol{b}}$ which, on the component X_1, is $\mathrm{Bl}_2 \, \mathbb{P}^2 \to \mathbb{P}^1 \times \mathbb{P}^1$. So, decreasing the weights does *not* give a morphism from $X_{\boldsymbol{b}}$ to $X_{\boldsymbol{b}'}$, unlike the curve case.

5.7.4 The case $n = 6$

Theorem 5.7.2 tells us precisely how to decode any partial tiling into a stable pair, for any weight $\boldsymbol{b}$. In Figures 5.12 and 5.13, we list the 25 stable pairs with weight $\boldsymbol{b} = \boldsymbol{1}$ for the 25 tilings of $\Delta(3,6)$ which we listed in Table 4.4, except for tiling number 7.

The stable pair for tiling number 7 is given in Figure 5.11. It is obtained by starting with a line arrangement of six lines meeting at four points, three at a time, blowing up the four points, and attaching to $\mathrm{Bl}_4\,\mathbb{P}^2$ four $\mathbb{P}^2$'s.

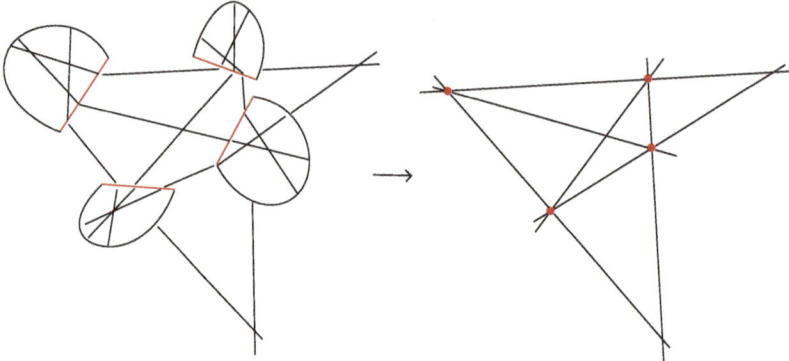

Figure 5.11: Stable pair for tiling number 7.

The other 24 stable pairs can be drawn in a toric way, so that the irreducible components correspond to polytopes forming a tiling of a triangle with side three. Recall that a dth multiple of the elementary triangle corresponds to $(\mathbb{P}^2, \mathcal{O}(d))$, a rhombus corresponds to $\mathbb{P}^1 \times \mathbb{P}^1$, and a trapezoid to $\mathrm{Bl}_1\,\mathbb{P}^2 = \mathbb{F}_1$. Thus the irreducible components of X are as in Table 5.1

no.	components
1, 5	$\mathrm{Bl}_1\,\mathbb{P}^2 + \mathbb{P}^2$
2, 3	$\mathrm{Bl}_2\,\mathbb{P}^2 + 2\,\mathbb{P}^2$
4	$\mathrm{Bl}_3\,\mathbb{P}^2 + 3\,\mathbb{P}^2$
6	$2\,\mathrm{Bl}_1\,\mathbb{P}^2 + \mathbb{P}^2$
7	$\mathrm{Bl}_4\,\mathbb{P}^2 + 4\,\mathbb{P}^2$
8	$\mathbb{P}^1 \times \mathbb{P}^1 + 2\,\mathbb{P}^2$
9, 10	$\mathbb{P}^1 \times \mathbb{P}^1 + \mathrm{Bl}_1\,\mathbb{P}^2 + 2\mathbb{P}^2$
11	$3\,\mathrm{Bl}_1\,\mathbb{P}^2$
12, 13, 14	$2\,\mathrm{Bl}_1\,\mathbb{P}^2 + \mathbb{P}^1 \times \mathbb{P}^1 + \mathbb{P}^2$
15-19	$\mathrm{Bl}_1\,\mathbb{P}^2 + 2\,\mathbb{P}^1 \times \mathbb{P}^1 + 2\,\mathbb{P}^2$
20-25	$3\,\mathbb{P}^1 \times \mathbb{P}^1 + 3\,\mathbb{P}^2$

Table 5.1: Irreducible components of stable hyperplane arrangements for $\boldsymbol{b} = 1$.

As we mentioned, all connected tilings in $\Delta(3,6)$ of codimension 1 can be extended to a complete tiling. So all weighted shas for $\boldsymbol{b} \neq \mathbf{1}$ are obtained from these by contraction and sometimes replacing $\mathbb{P}^1 \times \mathbb{P}^1$ by $\mathrm{Bl}_2 \, \mathbb{P}^2$, according to the rules of Theorem 5.7.2.

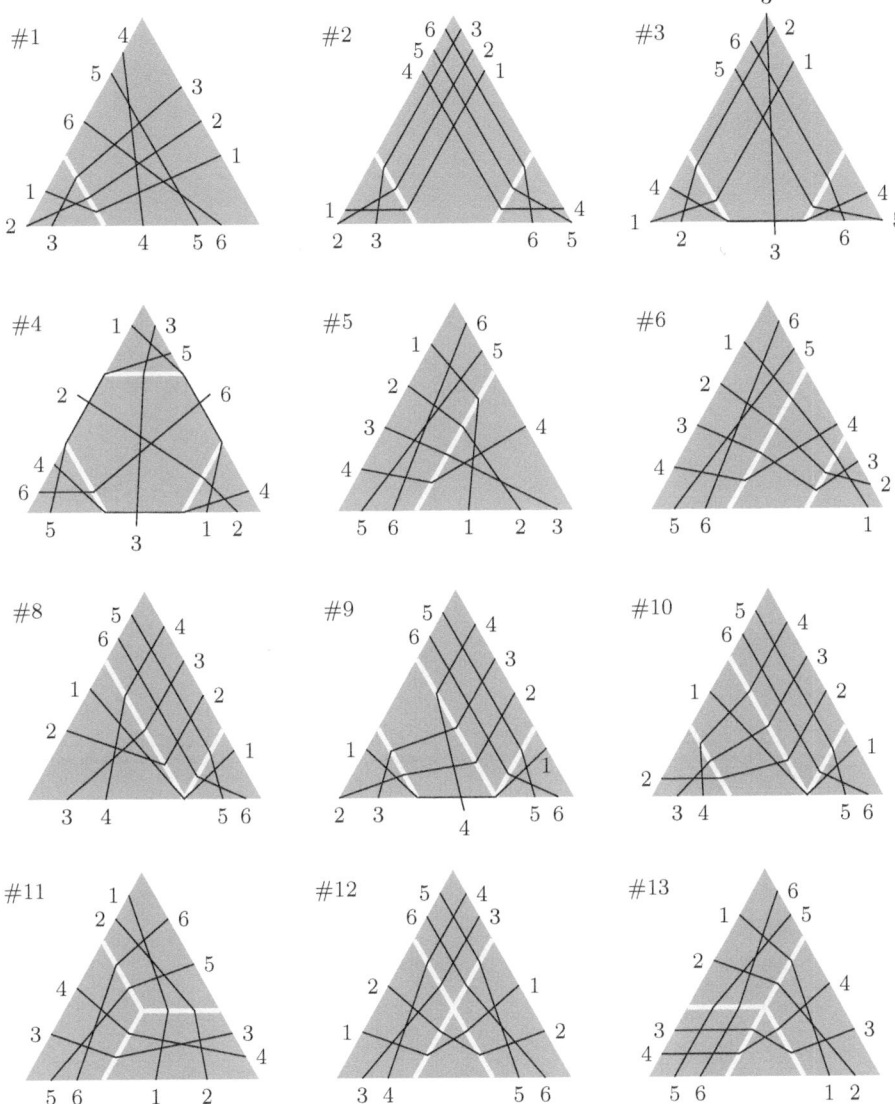

Figure 5.12: Shas of dimension 2 with $n = 6$.

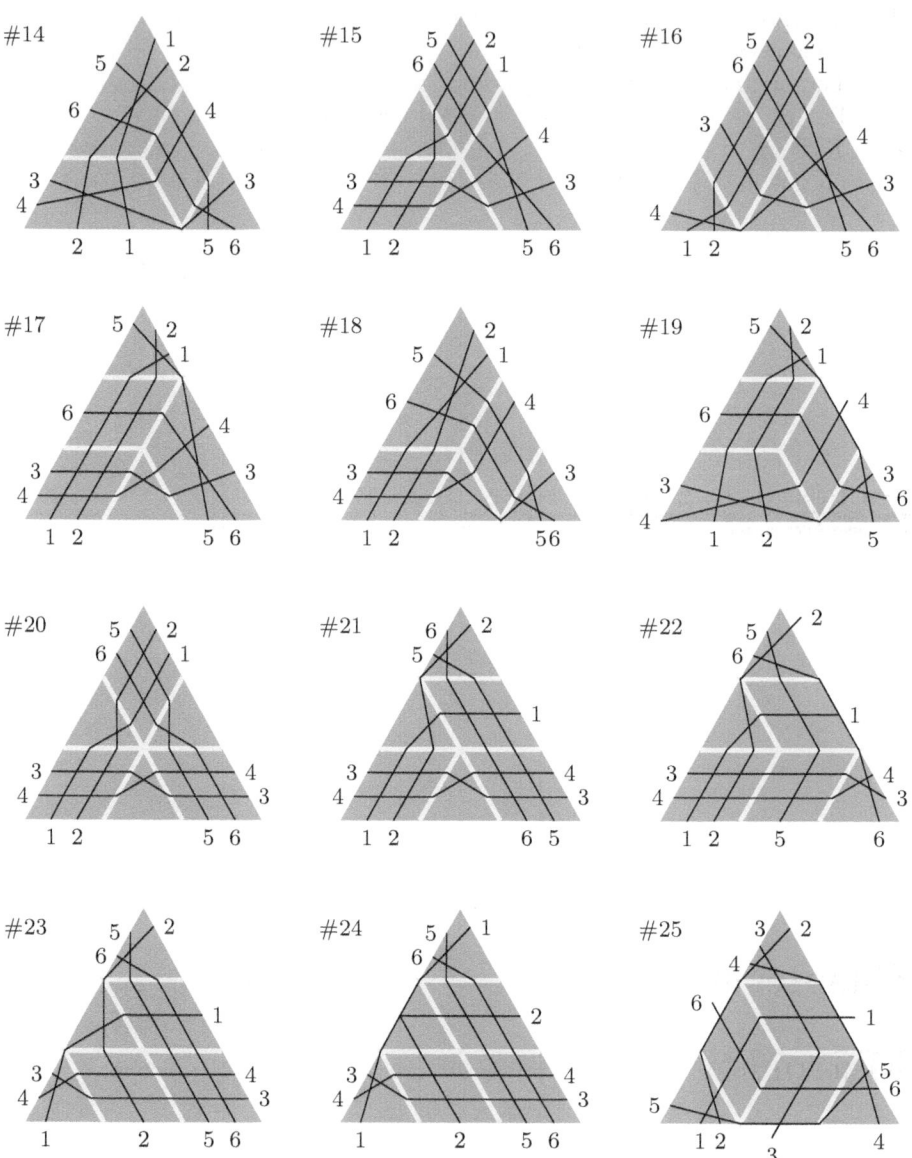

Figure 5.13: Shas of dimension 2 with $n = 6$, continued.

Chapter 6

Abelian Galois Covers

6.1 The yoga of cyclic and abelian Galois covers

6.1.1 Cyclic covers

Let $\pi\colon X \to Y$ be a cyclic Galois cover of two varieties for the group $G = \mu_n$ of roots of unity. Thus, we have a group action $\mu_n \curvearrowright X$ and the quotient is Y. Let us assume that X and Y are smooth for now, until we understand how to deal with the general case.

Remark 6.1.1. We may be tempted to write $G = \mathbb{Z}_n$, and we would be correct if we worked over $\mathbb{C}$ or over any field k of characteristic not dividing n which contains all n roots of unity. But that is only because over such a field the group varieties μ_n and $\mathbb{Z}_n$ are (non-canonically) isomorphic. It turns out that, in general, the μ_n-quotients are very simple, while $\mathbb{Z}_n$-quotients are very complicated and sometimes even pathological. So, let us do the correct and easy thing from the beginning.

The group of characters of G is $\mathbb{Z}_n$. Algebraically, this means that $\mu_n = \operatorname{Spec} k[\mathbb{Z}_n]$, where $k[\mathbb{Z}_n] = k[\lambda](\lambda^n - 1)$ is the group algebra of $\mathbb{Z}_n$. Now this formula works over any field. We can even use $k = \mathbb{Z}$ or any other ring R, and it would still work: we would get the group scheme $\mu_{n,R}$ over $\operatorname{Spec} R$. This is entirely similar to the case of a multiplicative group $\mathbb{G}_m = \operatorname{Spec} k[\mathbb{Z}]$ or a torus $T = \operatorname{Spec} k[M]$, $M \simeq \mathbb{Z}^n$, which can also be defined over any base.

Anyway, let us return back to the case when k is a nice field. How do we describe the cover π in terms of the data on the bottom variety Y? The morphism π is finite, and in particular affine. This means that $X = \operatorname{Spec}_Y \mathcal{A}$, where $\mathcal{A}$ is some $\mathcal{O}_Y$-algebra of rank n.

The μ_n-group action is the morphism, $G \times_Y X \to X$. Algebraically, this is described by a homomorphism of $\mathcal{O}_Y$-algebras, $\mathcal{A} \to \mathcal{A} \otimes k[\lambda]/(\lambda^n - 1)$. A pleasant and completely general argument shows this is equivalent to giving the algebra $\mathcal{A}$ a $\mathbb{Z}_n$-grading by the group of characters.

This is indeed a completely general fact, true over any base and for any *diagonalizable* group $\operatorname{Spec} k[G^*]$, where G^* is a finitely generated abelian group. The interested reader should read Grothendieck [18].

Thus, $\mathcal{A} = \mathcal{A}_0 \oplus \mathcal{A}_1 \oplus \cdots \oplus \mathcal{A}_{n-1}$. The $\mathbb{Z}_n$-action is described by the formula $\lambda \cdot (a_0, a_1, \ldots, a_{n-1}) = (a_0, \lambda a_1, \ldots, \lambda^{n-1} a_{n-1})$. The quotient is $\operatorname{Spec} A^G$, and the ring of invariants is obviously $\mathcal{A}_0$. Thus, one must have $\mathcal{A}_0 = \mathcal{O}_Y$.

The morphism π is flat. More generally, from commutative algebra we know that a finite R-module M over a regular ring R is flat over R if and only if M is Cohen–Macaulay. So, for as long as the bottom Y is smooth and the top X is Cohen–Macaulay, the sheaves $\mathcal{A}_i$ must be flat. A flat finite R-module is locally free. Since we also must have $\operatorname{rank} \mathcal{A}_i$, each $\mathcal{A}_i$ is an invertible $\mathcal{O}_Y$-module.

Finally, we should have the algebra structure on $\mathcal{A}$. This means that $\mathcal{A}_i$ can be identified with $\mathcal{A}_1^{\otimes i}$ and we must have the map $\mathcal{A}^{\otimes n} \to \mathcal{O}_Y$. Thus, the data for a μ_n-cover are the following: an invertible sheaf L on Y so that $\mathcal{A} = \mathcal{O}_Y \oplus L^{-1} \oplus L^{-2} \oplus \cdots \oplus L^{-(n-1)}$, and a homomorphism $L^{-n} \to \mathcal{O}_Y$, i.e., a section $s \in H^0(Y, L^n)$.

We still need to understand when such a cover X is smooth. This is easy: locally, $L \simeq \mathcal{O}_Y$ and s is a regular function on Y. The cover is locally given by the equation $z^n = s$. Thus, X is smooth if and only if the divisor $D = (s)$ is smooth.

When Y is projective and k has all n roots of unity, the section s can be replaced by any constant, and it is entirely determined by the divisor D. In this case, the data for the cover is an invertible sheaf L, and a smooth effective divisor D, which must satisfy the relation $L^n = D$. The divisor D is the ramification divisor of $\pi : X \to Y$.

6.1.2   Abelian Galois groups

A finite abelian group G is just a direct sum of several cyclic groups, hence a Galois G-cover $X \to Y$ can be decomposed as a sequence of cyclic covers. So, in principle, we trace them out using the previous subsection. However, it must be familiar to anyone that frequently things are nicer when you write formulas in a coordinate-free manner, without choosing a basis. This is the case here.

Also, even if X and Y are smooth, all the intermediate varieties are singular. Thus, the abelian covers do not really reduce to the case of cyclic covers of smooth varieties.

The general theory was described by Pardini in [52]. The data for a cover involves line bundles for all characters $\chi \in G^*$ and divisors $D_{H,\psi}$ for all cyclic subgroups $H \subset G$ and their generators ψ. This becomes quite cumbersome for a general group.

For a group of the form $G = \mu_p^n$ for a prime p, however, the pairs (H, ψ) are in bijection with the nontrivial elements of G. Since this is the only case we need, we will state the data for this case only. And it is very convenient to switch to the additive notation again, notwithstanding what was said above.

Thus, we fix $G \simeq \mathbb{Z}^p$, its dual group of characters $G^* \simeq \mathbb{Z}_p^p$, and a perfect pairing $G^* \times G \to \mathbb{Z}_p$, $(\chi, g) \mapsto \chi(g) \in \mathbb{Z}_p$. For a residue class $i \in \mathbb{Z}_p$, let $\bar{i} \in \{0, 1, \ldots, p-1\}$ be its smallest nonnegative lift to $\mathbb{Z}$.

Theorem 6.1.2. *The data for a Galois cover $\pi\colon X \to Y$ for the group $G = \mathbb{Z}_2^n$ are the following:*

1. *for each $\chi \in G^*$, an invertible line bundle L_g;*
2. *for each nonzero $g \in G$, an effective reduced divisor D_g (which could be zero);*

satisfying the fundamental relations (written here additively):

$$\forall \chi, \chi', \qquad L_\chi + L_{\chi'} = L_{\chi+\chi'} + \sum_{\chi(g)=\chi'(g)=1} D_g \quad in \ Pic(Y).$$

In particular, $2L_\chi = \sum_{\chi(g)=1} D_g$. One has $X = \mathrm{Spec}_Y \oplus_{\chi \in G^*} L_\chi^{-1}$.

Example 6.1.3. The data for a $\mathbb{Z}^2$ cover are three divisors A, B, C and three sheaves L_1, L_2, L_3 (plus $L_0 = \mathcal{O}_Y$) such that

$$L_1 + L_2 = L_3 + C, \qquad L_2 + L_3 = L_1 + A, \qquad L_3 + L_1 = L_2 + B,$$

which implies that

$$2L_1 = B + C, \qquad 2L_2 = C + A, \qquad 2L_3 = A + B.$$

Vice versa, if $\mathrm{Pic}(Y)$ has 2-torsion, the sheaves L_1, L_2, L_3 can be uniquely computed from A, B, C by the above formula.

Of course, the divisors should satisfy some local conditions that one can compute in order for X to be smooth.

Theorem 6.1.4 ([13]). *Assume that the group $\mathrm{Pic}\, Y$ has no torsion. Then a G-cover with the group $G = \mathbb{Z}_p^n$ for the ramification divisors D_g exists if and only if $\sum_g g D_g = 0$ in $\mathrm{Pic}(Y) \otimes \mathbb{Z}_p$. The line bundles L_χ can be computed uniquely from the divisors D_g by the formula*

$$p L_\chi = \sum_g \overline{\chi(g)} D_g.$$

6.1.3 Numerical invariants

Since $\mathcal{O}_X = \oplus_\chi L_\chi^{-1}$, one has

$$h^p(X, \mathcal{O}_X) = \sum_\chi h^p(Y, L_\chi^{-1}).$$

In the case $G = \mathbb{Z}_p^n$, the canonical class of X is computed by the formula

$$K_X = \pi^* \left(K_Y + \sum \frac{p-1}{p} D_g \right)$$

in $\mathrm{Pic}(X) \otimes \mathbb{Q}$, which is essentially the Riemann–Hurwitz formula for curves.

6.1.4   Singular covers

The work [14] extends this theory to the case of singular covers with at most double normal crossings in codimension 1. The most basic result here is the following theorem.

Theorem 6.1.5. *Suppose that $\pi\colon X \to Y$ is a finite cover of S_2 varieties with double crossings in codimension 1, B^X, B^Y are $\mathbb{Q}$-divisors on X, Y, and suppose also that, for the canonical divisors, the following formula holds:*

$$K_X + B^X = \pi^*\left(K_Y + B^Y\right).$$

Then (X, B^X) is slc if and only if (Y, B^Y) is slc. Also, $K_X + B^X$ is ample if and only if $K_Y + B^Y$ is ample.

How does it work can perhaps be guessed from the following example.

Example 6.1.6. Look at a double cover of surfaces $\pi\colon X \to Y$, defined by the data (L, D) such that $L^{\otimes 2} \simeq \mathcal{O}(D)$. Locally, it is given by the equation $z^2 = f(x, y)$, where f is a local equation for D.

When D is smooth, the cover is smooth. On the opposite side of the spectrum we have the case when $f(x, y) = x^2$, i.e., $D = 2E$ has a component of multiplicity 2. In this case, the local equation for X is $z^2 = x^2$. Therefore, X is non-normal, and has a double crossing singularity.

According to the above formula, $K_X = \pi^*(K_Y + \frac{1}{2}D)$. Since the pair $(Y, \frac{1}{2}D) = (Y, E)$ is lc, the surface X is slc.

6.2   Special K3 surfaces

6.2.1   Covers of $\mathbb{P}^2$ ramified in six lines

A polarized K3 surface is a pair (X, L), where X is a K3 surface, smooth or with ordinary double points, and L is an ample line bundle. The positive integer $L = 2d$ is always even, as the intersection form on a K3 surface is even.

The smallest possible degree is 2. A K3 surface of degree 2 is a double cover $\pi\colon X \to Y$, where Y is either $\mathbb{P}^2$ or $\mathbb{F}_4^0$, a surface obtained from the Hirzebruch surface $\mathbb{F}_4$ by contracting the (-4)-section; it is a cone over the rational quartic curve in $\mathbb{P}^4$.

By the adjunction formula, $K_X = \pi^*(K_Y + \frac{1}{2}D)$. The branch divisor D thus satisfies $\frac{1}{2}D \sim_{\mathbb{Q}} -K_X$. Therefore, it is a sextic curve in the case $Y = \mathbb{P}^2$.

When C_d is a curve of degree d, the Hacking compactification of the planar pairs looks at the pairs $(\mathbb{P}^2, (\frac{3}{d} + \epsilon)C_d)$. For sextics, this gives $(\mathbb{P}^2, (\frac{1}{2} + \epsilon)D)$.

While it is hard to do it for the whole 19-dimensional family of all sextic curves, let us treat the case when D is a union of six lines. In terms of weighted hyperplane arrangements, this is the quotient of the moduli space $M_b(3, 6)$, where

$\boldsymbol{b} = (\frac{1}{2}+\epsilon, \ldots, \frac{1}{2}+\epsilon)$, divided by the symmetric group S_6. Recall that $\dim M_b(3,6) = (6-4)(3-1) = 4$.

For this space, we have the compactification $\overline{M}_b(3,6)$ which, on the boundary, adds weighted shas. By Theorem 6.1.5, the double covers $(X, \epsilon R)$ are stable pairs, where $R = \pi^{-1}(B) = \frac{1}{2}\pi^*(B)$ is the ramification divisor.

According to our recipe, to compute the stable pairs we have to compute all matroid covers of the $\boldsymbol{b}$-cut polytope $\Delta_b(3,6)$. This is a very small polytope for $0 < \epsilon \ll 1$, a small neighborhood of the central point $(\frac{1}{2}, \ldots, \frac{1}{2})$ of $\Delta(3,6)$.

Since we already computed all complete matroid covers of $\Delta(3,6)$ and we already stated, without proof, that any partial cover (connected in codimension 1) extends to a complete cover, all we have to do is to look at the neighborhood of the central point in the lists from Figures 5.11, 5.12, and 5.13.

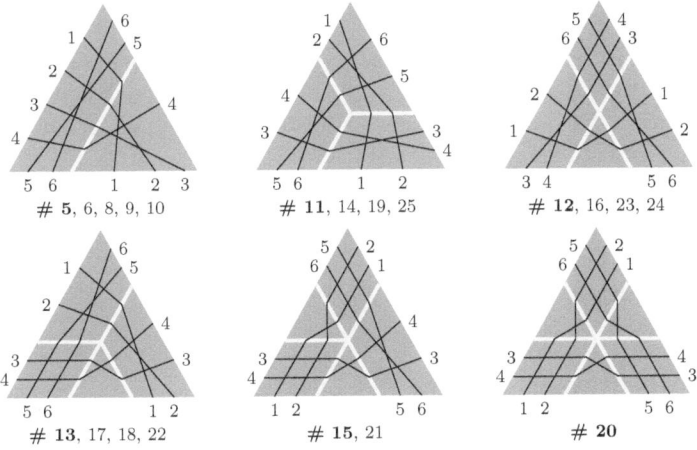

Figure 6.1: Degenerations of some K3 surfaces of degree 2.

For many of them the central point lies in the interior of one of the polytopes BP_V. In that case, the matroid cover of Δ_b consists of the single polytope BP_V, the $\mathbb{P}^2$ does not degenerate, the cover X is normal, and the pair $(X, \epsilon R)$ is log canonical.

The smaller-dimensional polytopes containing the point $a_0 = (\frac{1}{2}, \ldots, \frac{1}{2})$ have faces of the form $x_{12} \leq 1$, respectively $x_{3456} \leq 2$. The codimension 2 polytopes containing a_0 are of the form $x_{12} = x_{34} = x_{56} = 1$, plus all the S_6 permutations, of course.

The resulting pairs are listed in Figure 6.1. There are six nontrivial cases, in addition to the trivial case $Y = \mathbb{P}^2$. The numbers under the picture are all the types from Figures 5.11, 5.12, 5.13 that produce this weighted sha. The circled number is the easiest type.

One should note that if we did not have the complete tilings of $\Delta(3,6)$, then we would not need them all. For our computation, we only need to look at the polytopes intersecting a very small neighborhood of the center a.

Using the formulas of Subsection 6.1.3, it is easy to compute the irreducible components Z of the degenerate K3 surfaces. Note that, since the sums of the B_i's on each components are divisible by 2, the components of the double locus D are *not* in the branch locus.

The cover of a component $\mathbb{P}^2$ in Picture no. 1, ramified in four lines, has

$$K_Z = \pi^*(K_{\mathbb{P}^2} + \frac{1}{2}(4h)) = \pi^*(-h).$$

Therefore, Z is a del Pezzo surface with $K_Z^2 = \pi^*(h^2) = 2$. A generic such surface has six singularities of type A_1 with equation $z^2 = xy$ over the six intersection points of the four lines.

The cover of $\mathbb{F}_1$ in Picture 1, ramified in $2(s_1 + f) + 4f$, has $K_Z = \pi^*(-s_1)$. Therefore, $K_{X_1}^2 = -2$. One has $K_X = -R$, and R is an elliptic curve with $R^2 = -2$.

The cover of $\mathbb{F}_1$ in Pictures 2, 3, 4, and 5, ramified in $2(s_1 + f) + 2f$, has $K_Z = \pi^*(-s_1 - f)$. Therefore, $K_Z^2 = 2$ and $-K_Z$ is semiample and contracts a (-2)-curve to a point. The surface Z is a partial resolution of an A_1 singularity on a del Pezzo surface of degree two with six A_1 points.

The cover of $\mathbb{P}^1 \times \mathbb{P}^1$ in Pictures 3, 4, 5, and 6 is a del Pezzo surface of degree four with four A_1 points.

Finally, the cover of $\mathbb{P}^2$ in Pictures 3, 4, 5, and 6 is a del Pezzo surface of degree eight with a single A_1 singularity.

6.2.2   Degenerations of Kummer surfaces

To every abelian surface A one can associate its Kummer surface $X = A/\langle \pm 1 \rangle$. It has sixteen ordinary nodes. Now suppose that (A, L) is a principally polarized surface. Then the sheaf $L \otimes (-1)^* L$ descends to X and realizes X as a quartic surface in $\mathbb{P}^3$ with sixteen nodes. Projecting from one of the nodes gives a double cover $\pi: X \to \mathbb{P}^2$ ramified in six lines; the remaining fifteen nodes are the π-preimages of the $\binom{6}{2}$ points of intersections of these six lines.

Thus, the Kummers of principally polarized abelian surfaces form a closed three-dimensional subfamily of the four-dimensional family $M \subset M_b(3, 6)$ from the previous subsection. They are distinguished by the condition that the six lines $B_1, \dots, B_6$ are tangent to a common conic C, at the points $P_1, \dots, P_6$.

Let $\widetilde{C} \to C$ be the 2-to-1 cover ramified in $P_1, \dots, P_6$. Then $\widetilde{C}$ is a curve of genus 2 and the abelian surface A is its Jacobian, $J\widetilde{C}$.

The degenerations of the Jacobians are very well understood (see [5]) and the degenerations of their Kummers are intimately related to them. Among the six degenerations of the previous section, only three appear: cases 5, 12, and 20. They correspond to the degenerations in which 1, 2, or 3 of the pairs of points among the points $P_1, \dots, P_6$ come together.

6.3 Numerical Campedelli surfaces

This case is taken from [13]. Numerical Campedelli surfaces that we consider are $\mathbb{Z}_2^3$-Galois covers of $\mathbb{P}^2$ with D_g being a line for each $g \neq 0$. The adjunction formula says

$$K_X = \pi^*\left(K_{\mathbb{P}^2} + \frac{1}{2}(7h)\right).$$

Therefore, $K_X^2 = 8 \times 1/4 = 2$, and K_X is ample. Thus, X is a surface of general type with $K_X^2 = 2$. One further computes that

$$p_g = h^2(\mathcal{O}_X) = 0 \quad \text{and} \quad q = h^1(\mathcal{O}_X) = 0.$$

The moduli space of such surfaces is $M_b(3,7)$ for $\boldsymbol{b} = (\frac{1}{2}, \dots, \frac{1}{2})$, and its compactification is $\overline{M}_b(3,7)$. To compute the latter, we need to compute the matroid covers of $\Delta_b(3,7)$. However, all such matroid covers are trivial. Indeed, none of the hyperplanes $\boldsymbol{x}(I) = 2$, i.e., $\boldsymbol{x}(I^c) = 1$, intersect the interior of $\Delta_b(3,7)$. If all $x_i \leq 1/2$, then $|I| = 6$, but then $|I^c| = 1$ and $\boldsymbol{x}(I^c) \leq 1/2$.

Therefore, $M_b(3,7) = \overline{M}_b(3,7)$ is already compact and equal to the GIT quotient $G(3,7)/\!/_b T$ for the symmetric ("democratic") weight.

6.4 Kulikov surfaces

Kulikov surfaces are $\mathbb{Z}_3^2$-covers of $\mathrm{Bl}_3\,\mathbb{P}^2$ in a configuration of nine curves obtained by blowing the vertices of the triangle in the configuration of six lines in $\mathbb{P}^2$ pictured in Figure 6.2. The colors of the divisors D_g correspond to the following group

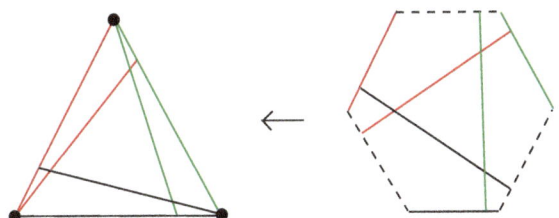

Figure 6.2: Kulikov surface configuration.

elements $g \in \mathbb{Z}_3^2$: red = $(1,0)$, green = $(1,1)$, black = $(1,2)$, and the dashed line is $(0,1)$. These surfaces are smooth, have ample K_X, $K_X^2 = 6$, and $p_g = q = 1$.

The moduli space together with its compactification was considered in [16], from a different point of view. We give it here to illustrate our methods.

It is fairly obvious that the configuration of six lines forms a one-dimensional family. So, compactifying it should not be too hard.

The starting configuration is that of tiling number 4 in Figure 5.12. The subdivisions must have their corners "cut off". The only such subdivision is number 25 in Figure 5.13. The tilings in Figure 5.13 are given modulo S_6, so in fact there are two degenerations, shown in Figure 6.3.

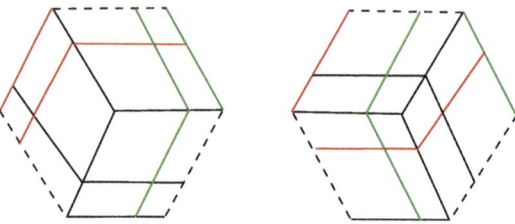

Figure 6.3: Degenerations of Kulikov surfaces.

The compactified moduli space is isomorphic to $\mathbb{P}^1$.

Bibliography

[1] V. Alexeev. *Boundedness and K^2 for log surfaces*, Internat. J. Math. **5** (1994), no. 6, 779–810.

[2] V. Alexeev. *Log canonical singularities and complete moduli of stable pairs.* `arXiv:alg-geom/9608013(1996)`.

[3] V. Alexeev. *Moduli spaces $M_{g,n}(W)$ for surfaces*, Higher-dimensional Complex Varieties (Trento, 1994), de Gruyter, Berlin, 1996, pp. 1–22.

[4] V. Alexeev. *Complete moduli in the presence of semiabelian group action*, Ann. of Math. (2) **155** (2002), no. 3, 611–708.

[5] V. Alexeev. *Compactified Jacobians and Torelli map*, Publ. Res. Inst. Math. Sci. **40** (2004), no. 4, 1241–1265.

[6] V. Alexeev. *Limits of stable pairs*, Pure Appl. Math. Q. **4** (2008), no. 3, Special Issue: In honor of Fedor Bogomolov. Part 2, 767–783. `arXiv: math/0607684(2006)`.

[7] V. Alexeev. *Weighted stable hyperplane arrangements.* `arXiv: 0806.0881(2008)`.

[8] V. Alexeev and M. Brion. *Stable reductive varieties. II. Projective case*, Adv. Math. **184** (2004), no. 2, 380–408.

[9] V. Alexeev and M. Brion. *Stable spherical varieties and their moduli*, IMRP Int. Math. Res. Pap. (2006), Art. ID 46293, 57.

[10] V. Alexeev and A. Brunyate. *Extending the Torelli map to toroidal compactifications of Siegel space*, Invent. Math. **188** (2012), no. 1, 175–196.

[11] V. Alexeev and G.M. Guy. *Moduli of weighted stable maps and their gravitational descendants*, J. Inst. Math. Jussieu **7** (2008), no. 3, 425–456.

[12] V. Alexeev and A. Knutson. *Complete moduli spaces of branchvarieties*, J. Reine Angew. Math. **639** (2010), 39–71. `arXiv: math/0602626(2006)`.

[13] V. Alexeev and R. Pardini. *Explicit compactifications of moduli spaces of Campedelli and Burniat surfaces.* Preprint (2009), 26 pp. `arXiv:0901.4431(2009)`.

[14] V. Alexeev and R. Pardini. *Non-normal abelian covers*, Compositio Math. **148** (2012), 1051–1084. `arXiv:1102.4184(2011)`.

[15] C. Birkar, P. Cascini, C.D. Hacon and J. McKernan. *Existence of minimal models for varieties of log general type*, J. Amer. Math. Soc. **23** (2010), no. 2, 405–468.

[16] T.O.M. Chan and S. Coughlan. *Kulikov surfaces form a connected component of the moduli space*, Nagoya Math. J. **210** (2013), 1–27. arXiv:1011.5574(2012).

[17] P. Deligne and D.B. Mumford. *The irreducibility of the space of curves of given genus*, Inst. Hautes Études Sci. Publ. Math. (1969), no. 36, 75–109.

[18] M. Demazure, A. Grothendieck. *SGA3. Schémas en groupes*, vol. 151, 152, 153, (1970).

[19] I.V. Dolgachev and Y. Hu. *Variation of geometric invariant theory quotients*, Inst. Hautes Études Sci. Publ. Math. (1998), no. 87, 5–56, with an appendix by Nicolas Ressayre.

[20] O. Fujino. *Semipositivity theorems for moduli problems.* arXiv:1210.5784(2012).

[21] W. Fulton. *Introduction to Toric Varieties*, Annals of Mathematics Studies, vol. 131, Princeton University Press, Princeton, NJ, 1993.

[22] I.M. Gel'fand, R.M. Goresky, R.D. MacPherson, and V.V. Serganova. *Combinatorial geometries, convex polyhedra and Schubert cells*, Adv. in Math. **63** (1987), no. 3, 301–316.

[23] I.M. Gel'fand, M.M. Kapranov and A.V. Zelevinsky. *Discriminants, Resultants, and Multidimensional Determinants*, Mathematics: Theory & Applications, Birkhäuser Boston Inc., Boston, MA, 1994.

[24] I.M. Gel'fand and V.V. Serganova. *Combinatorial geometries and the strata of a torus on homogeneous compact manifolds*, Uspekhi Mat. Nauk **42** (1987), no. 2(254), 107–134, 287.

[25] T. Gwena. *Degenerations of cubic threefolds and matroids*, Proc. Amer. Math. Soc. **133** (2005), no. 5, 1317–1323 (electronic).

[26] P. Hacking. *Compact moduli of plane curves*, Duke Math. J. **124** (2004), no. 2, 213–257.

[27] P. Hacking, S. Keel and J. Tevelev. *Compactification of the moduli space of hyperplane arrangements*, J. Algebraic Geom. **15** (2006), no. 4, 657–680.

[28] P. Hacking, S.K. and J. Tevelev. *Stable pair, tropical, and log canonical compactifications of moduli spaces of del Pezzo surfaces*, Invent. Math. **178** (2009), no. 1, 173–227.

[29] C.D. Hacon and C. Xu. *Existence of log canonical closures*, Invent. Math. **192** (2013), no. 1, 161–195.

[30] B. Hassett. *Moduli spaces of weighted pointed stable curves*, Adv. Math. **173** (2003), no. 2, 316–352.

[31] B. Hassett and S.J. Kovács. *Reflexive pull-backs and base extension*, J. Algebraic Geom. **13** (2004), no. 2, 233–247.

[32] S. Herrmann, A. Jensen, M. Joswig and B. Sturmfels. *How to draw tropical planes*, Electron. J. Combin. **16** (2009), no. 2, Special volume in honor of Anders Bjorner, Research Paper 6, 26.

[33] M.M. Kapranov. *Chow quotients of Grassmannians I*, Izrail' M. Gel'fand Seminar, Adv. Soviet Math., vol. 16, Amer. Math. Soc., Providence, RI, 1993, pp. 29–110.

[34] M. Kawakita. *Inversion of adjunction on log canonicity*, Invent. Math. **167** (2007), no. 1, 129–133.

[35] S. Keel and S. Mori. *Quotients by groupoids*, Ann. of Math. (2) **145** (1997), no. 1, 193–213.

[36] J. Kollár. *Projectivity of complete moduli*, J. Differential Geom. **32** (1990), no. 1, 235–268.

[37] J. Kollár. *Quotient spaces modulo algebraic groups*, Ann. of Math. (2) **145** (1997), no. 1, 33–79.

[38] J. Kollár. *Hulls and husks*. arXiv:0805.0576(2008).

[39] J. Kollár. *Moduli of varieties of general type*. arXiv:1008.0621(2010).

[40] J. Kollár. *Sources of log canonical centers*. arXiv:1107.2863(2011).

[41] J. Kollár. *Book on Moduli of Surfaces, work in progress*, Available at https://web.math.princeton.edu/ kollar/.

[42] J. Kollár and N.I. Shepherd-Barron. *Threefolds and deformations of surface singularities*, Invent. Math. **91** (1988), no. 2, 299–338.

[43] R. Laza. *The KSBA compactification for the moduli space of degree two K3 pairs*. arXiv:1205.3144(2012).

[44] W. Liu. *Stable degenerations of surfaces isogenous to a product II*, Trans. Amer. Math. Soc. **364** (2012), no. 5, 2411–2427.

[45] Y. Matsumoto, S. Moriyama, H. Imai and D. Bremner. *Matroid enumeration for incidence geometry*, Discrete Comput. Geom. **47** (2012), no. 1, 17–43.

[46] M. Melo and F. Viviani, *Comparing perfect and 2nd Voronoi decompositions: the matroidal locus*, Math. Ann. **354** (2012), no. 4, 1521–1554.

[47] D. Mumford, J. Fogarty and F. Kirwan. *Geometric Invariant Theory*, third ed., Ergebnisse der Mathematik und ihrer Grenzgebiete (2) [Results in Mathematics and Related Areas (2)], vol. 34, Springer-Verlag, Berlin, 1994.

[48] T. Oda. *Convex Bodies and Algebraic Geometry*, Ergebnisse der Mathematik und ihrer Grenzgebiete (3), vol. 15, Springer-Verlag, Berlin, 1988.

[49] Y. Odaka. *The git stability of polarized varieties via discrepancy*, Annals of Mathematics **177** (2013), no. 2, 645–661.

[50] J.G. Oxley. *Matroid Theory*, Oxford Science Publications, The Clarendon Press Oxford University Press, New York, 1992.

[51] J. Oxley, *What is a matroid?*, Cubo Mat. Educ. **5** (2003), no. 3, 179–218.

[52] R. Pardini. *Abelian covers of algebraic varieties*, J. Reine Angew. Math. **417** (1991), 191–213.

[53] A. Schrijver. *Combinatorial Optimization. Polyhedra and Efficiency. Vol. B*, Algorithms and Combinatorics, vol. 24, Springer-Verlag, Berlin, 2003, Matroids, trees, stable sets, Chapters 39–69.

[54] P.D. Seymour. *Decomposition of regular matroids*, J. Combin. Theory Ser. B**28** (1980), no. 3, 305–359.

[55] I.R. Shafarevich. *Lectures on Minimal Models and Birational Transformations of Two-dimensional Schemes*, Notes by C.P. Ramanujam. Tata Institute of Fundamental Research Lectures on Mathematics and Physics, no. 37, Tata Institute of Fundamental Research, Bombay, 1966.

[56] D. Speyer and B. Sturmfels. *The tropical Grassmannian*, Adv. Geom. **4** (2004), no. 3, 389–411.

[57] R. Vakil. *Murphy's law in algebraic geometry: badly-behaved deformation spaces*, Invent. Math. **164** (2006), no. 3, 569–590.

[58] M.A. van Opstall. *Moduli of products of curves*, Arch. Math. (Basel) **84** (2005), no. 2, 148–154.

[59] M.A. van Opstall. *Stable degenerations of surfaces isogenous to a product of curves*, Proc. Amer. Math. Soc. **134** (2006), no. 10, 2801–2806 (electronic).

[60] M.A. van Opstall. *Stable degenerations of symmetric squares of curves*, Manuscripta Math. **119** (2006), no. 1, 115–127.

[61] E. Viehweg. *Quasi-projective Moduli for Polarized Manifolds*, Ergebnisse der Mathematik und ihrer Grenzgebiete (3) [Results in Mathematics and Related Areas (3)], vol. 30, Springer-Verlag, Berlin, 1995.

[62] N.L. White. *The basis monomial ring of a matroid*, Advances in Math. **24** (1977), no. 3, 292–297.

[63] N. White (ed.). *Theory of Matroids*, Encyclopedia of Mathematics and its Applications, vol. 26, Cambridge University Press, Cambridge, 1986.

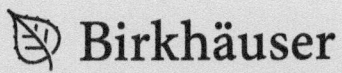 **Birkhäuser** | **birkhauser-science.com**

Advanced Courses in Mathematics – CRM Barcelona (ACM)

Edited by
Enric Ventura, Universitat Politècnica de Catalunya

Since 1995 the Centre de Recerca Matemàtica (CRM) has organised a number of Advanced Courses at the post-doctoral or advanced graduate level on forefront research topics in Barcelona. The books in this series contain revised and expanded versions of the material presented by the authors in their lectures.

■ **Dai, F. / Xu, Y.,** Analysis on *h*-Harmonics and Dunkl Transforms (2015).
ISBN 978-3-0348-0886-6
As a unique case in this Advanced Courses book series, the authors have jointly written this introduction to *h*-harmonics and Dunkl transforms. These notes provide an overview of what has been developed so far. The first chapter gives a brief recount of the basics of ordinary spherical harmonics and the Fourier transform. The Dunkl operators, the intertwining operators between partial derivatives and the Dunkl operators are introduced and discussed in the second chapter. The next three chapters are devoted to analysis on the sphere, and the final two chapters to the Dunkl transform. The authors' focus is on the analysis side of both h-harmonics and Dunkl transforms. The need for background knowledge on reflection groups is kept to a bare minimum.

■ **Citti, G. / Grafakos, L. / Pérez, C. / Sarti, A. / Zhong, X.,** Harmonic and Geometric Analysis (2015).
ISBN 978-3-0348-0407-3
This book presents an expanded version of four series of lectures delivered by the authors at the CRM. The first lecture is an application of harmonic analysis and the Heisenberg group to understanding human vision, while the second and third series of lectures cover some of the main topics on linear and multilinear harmonic analysis. The last serves as a comprehensive introduction to a deep result from De Giorgi, Moser and Nash on the regularity of elliptic partial differential equations in divergence form.

■ **Böckle, G. / Burns, D. / Goss, D. / Thakur, D. / Trihan, F. / Ulmer, D.,** Arithmetic Geometry over Global Function Fields (2014).
ISBN 978-3-0348-0852-1
This volume collects the texts of five courses given in the Arithmetic Geometry Research Programme 2009–2010 at the CRM Barcelona. All of them deal with characteristic *p* global fields; the common theme around which they are centered is the arithmetic of *L*-functions (and other special functions), investigated in various aspects. Three courses examine some of the most important recent ideas in the positive characteristic

theory discovered by Goss (a field in tumultuous development, which is seeing a number of spectacular advances): they cover respectively crystals over function fields (with a number of applications to *L*-functions of *t*-motives), gamma and zeta functions in characteristic *p*, and the binomial theorem. The other two are focused on topics closer to the classical theory of abelian varieties over number fields: they give respectively a thorough introduction to the arithmetic of Jacobians over function fields (including the current status of the BSD conjecture and its geometric analogues, and the construction of Mordell-Weil groups of high rank) and a state of the art survey of Geometric Iwasawa Theory explaining the recent proofs of various versions of the Main Conjecture, in the commutative and non-commutative settings.

■ **Asaoka, M. / El Kacimi Alaoui, A. / Hurder, S. / Richardson, K.,** Foliations: Dynamics, Geometry and Topology (2014).
ISBN 978-3-0348-0870-5
The lectures by A. El Kacimi Alaoui offer an introduction to foliation theory, with emphasis on examples and transverse structures. S. Hurder's lectures apply ideas from smooth dynamical systems to develop useful concepts in the study of foliations, like limit sets and cycles for leaves, leafwise geodesic flow, transverse exponents, stable manifolds, Pesin theory, and hyperbolic, parabolic, and elliptic types of foliations, all of them illustrated with examples. The lectures by M. Asaoka are devoted to the computation of the leafwise cohomology of orbit foliations given by locally free actions of certain Lie groups, and its application to the description of the deformation of those actions. In the lectures by K. Richardson, he studies the geometric and analytic properties of transverse Dirac operators for Riemannian foliations and compact Lie group actions, and explains a recently proved index formula.

■ **Alesker, S. / Fu, J. H. G.,** Integral Geometry and Valuations (2014).
ISBN 978-3-0348-0873-6

■ **Cruz-Uribe, D. / Fiorenza, A. / Ruzhansky, M. / Wirth, J.,** Variable Lebesgue Spaces and Hyperbolic Systems (2014).
ISBN 978-3-0348-0839-2